Multihomed Communication with **SCTP**

(Stream Control Transmission Protocol)

OTHER TELECOMMUNICATIONS BOOKS FROM AUERBACH

AUERBACH PUBLICATIONS
www.auerbach-publications.com
To Order Call: 1-800-272-7737 • Fax: 1-800-374-3401
E-mail: orders@crcpress.com

Multihomed
Communication with **SCTP**

(Stream Control Transmission Protocol)

Edited by Victor C.M. Leung,
Eduardo Parente Ribeiro,
Alan Wagner, and Janardhan Iyengar

CRC Press
Taylor & Francis Group
Boca Raton London New York

CRC Press is an imprint of the
Taylor & Francis Group, an **informa** business

CRC Press
Taylor & Francis Group
6000 Broken Sound Parkway NW, Suite 300
Boca Raton, FL 33487-2742

First issued in paperback 2016

© 2013 by Taylor & Francis Group, LLC
CRC Press is an imprint of Taylor & Francis Group, an Informa business

No claim to original U.S. Government works

Version Date: 2012918

ISBN 13: 978-1-138-03372-6 (pbk)
ISBN 13: 978-1-4665-6698-9 (hbk)

Library of Congress Cataloging-in-Publication Data

Multihomed communication with SCTP (Stream Control Transmission Protocol) /
editors, Victor Leung ... [et al.].
    p. cm.
Includes bibliographical references and index.
ISBN 978-1-4665-6698-9 (hardback)
    1. Stream Control Transmission Protocol (Computer network protocol) I. Leung,
Victor Chung Ming, 1955-

TK5105.5835.M85 2012
004.6--dc23   2012031866

Visit the Taylor & Francis Web site at
http://www.taylorandfrancis.com

and the CRC Press Web site at
http://www.crcpress.com

Contents

Preface

SCTP (Stream Control Transmission Protocol) is a relatively new standards track transport layer protocol in the Internet Engineering Task Force (IETF) described in RFC4960. While SCTP was originally intended for telephony signalling over IP, it was recognized and designed as a general purpose transport protocol for the Internet. Like Transmission Control Protocol (TCP), SCTP offers a reliable, full-duplex connection with mechanisms for flow and congestion control. Unlike both TCP and User Datagram Protocol (UDP), SCTP offers new delivery options and several other features and services. This book focuses on SCTP multihoming, an innovative feature that allows a transport layer association to span multiple IP addresses at each endpoint. SCTP multihoming allows an endpoint to simultaneously maintain and use multiple points of connectivity to the network; thus, fixed and mobile users could connect to the Internet via multiple service providers and/or last hop technologies, and could use one or potentially all of those connections.

List of Contributors

Armando Caro
BBN Technologies
Cambridge, Massachusetts
United States

Claudio Casetti
Politecnico di Torino
Torino, Italy

Lode Coene
Eandis Delaware Consulting
Antwerp, Belgium

Janardhan Iyengar
Franklin & Marshall College
Lancaster, Pennsylvania
United States

Victor C. M. Leung
University of British Columbia
Vancouver, British Columbia
Canada

Li Ma
University of British Columbia
Vancouver, British Columbia
Canada

Eduardo Parente Ribeiro
Federal University of Paraná
Curitiba, Paraná
Brazil

Brad Penoff
Google, Inc.
Mountain View, California
United States

Irene Rüngeler
Münster University of Applied Sciences
Münster, Germany

Michael Tüxen
Münster University of Applied Sciences
Münster, Germany

Alan Wagner
University of British Columbia
Vancouver, British Columbia
Canada

F. Richard Yu
Carleton University
Ottawa, Ontario
Canada

1

Fundamental Concepts and Mechanisms of Stream Control Transmission Protocol (SCTP) Multihoming

Claudio Casetti

The SCTP protocol was initially introduced by the Internet Engineering Task Force (IETF) as a transport layer protocol for the SS7 signalling network. Since then, SCTP has attracted increased attention as a de-facto member of the TCP/IP suite. Complementing some of the deficiencies of both Transmission Control Protocol (TCP) and User Datagram Protocol (UDP), the SCTP protocol offers advanced, more secure connection set-ups, data handling through multiple logical streams, as well as fully and partially reliable delivery. Arguably, the most appealing feature of SCTP is its support of multihoming for hosts with multiple network interfaces. In this introductory chapter, we will provide a classification and review of the research work that, in the decade since its introduction, has focused on the many facets of SCTP multihoming.

1.1   Introducing SCTP

The origins of SCTP are to be traced back to the attempts at merging the traditional telephone network, the Public Switched Telephone Network (PSTN), and IP networks. Indeed, combining the ubiquity of IP networks with the superior resource utilization provided by packet switching (as opposed to circuit switching), seemed to be a sure recipe for success, something later testified by the widespread diffusion of applications such as Skype or Google Voice. Such marriage, which came to be known as Voice over IP (VoIP), ran into early troubles. Predictably, the major hurdle was the Quality of Service (QoS) experienced by VoIP calls transported over the IP network, whose service model only provides best effort delivery. This shortcoming has been traditionally solved by "throwing bandwidth at the problem," i.e., by over-provisioning networks thereby avoiding congestion (and letting TCP bear the brunt of it when it arose). Additionally, true integration between VoIP and PSTN could only be achieved if the signaling network used by the latter, called SS7, could interoperate with both traditional phone sets and the upcoming VoIP devices. Beside allowing the setting up and managing of VoIP calls, an IP/SS7 integration could help blending in multimedia and other advanced applications. To this end, in 1998 the IETF created the Signaling Transport (SIGTRAN) working group. Its aim, as stated in its charter (Sigtran 2007), was to "address the transport of packet-based PSTN signaling over IP networks, taking into account functional and performance requirements of the PSTN signaling."

SIGTRAN initially pursued its goal by translating existing telephone signaling protocols into new ones that could operate on a pure IP-based network. The requirements of these signaling protocols, chiefly loss/delay bounds, security and resiliance, were to be handled by the transport layer and the first attempt was to use the only widely available reliable transport-layer protocols, i.e., TCP. However, TCP had some shortcomings that prevented its use (Stewart 2007):

- offering an in-order delivery, TCP suffers from Head-of-the-Line (HoL) blocking and could unnecessarily delay application data that need reliable transfer without sequence maintenance;

- its stream-oriented nature forces applications to delineate message boundaries;

- TCP is vulnerable to Denial-of-Service (DoS) attacks, such as SYN floods;

- in case of multihomed hosts, TCP sockets are not flexible enough to handle hot switching between network interfaces or to fully exploit parallel transmissions on multiple interfaces.

Overcoming the above limitations for PSTN signalling, and, nominally, for other applications that had similar needs, was the driving force pushing for the design and development of a new transport protocol. The result, RFC 4960 claims, was a protocol that was "designed to transport PSTN signaling messages over IP networks," but was "capable of broader applications" (Stewart 2007). SCTP was born.

1.2 SCTP Overview

Like all transport protocols in the so-called Internet Suite, SCTP (Stream Control Transport Protocol) sits atop the IP layer, as shown in Figure 1.1, and complements the IP service model: sharing many features of TCP, it offers a point-to-point, connection-oriented, reliable delivery service. Furthermore, it inherits the flow and congestion control algorithms of TCP, a crucial requirements that guarantees the "friendliness" of the new protocol toward current TCP stacks. Besides these similarities, however, TCP and SCTP can be considered distant relatives, the main differences being:

- SCTP creates an *association* between two endpoints, as opposed to a TCP connection: the two relationships differ in terms of features, data handling and security options;

- TCP handles data in a single stream, whereas SCTP can set up *multiple logical streams* that allow advanced delivery options;

- TCP is a stream-oriented protocol, while SCTP is *message-oriented*;

- both TCP and SCTP offer reliable delivery of all data, but SCTP has the additional option of *partial reliability* to expedite reception of time-constraint data;

- a TCP connection is bound to a single network interface, while SCTP supports *multihoming*.

In the rest of the section, we will provide a brief overview of the above differences, except for multihoming, which will be discussed in detail in Section 1.3.

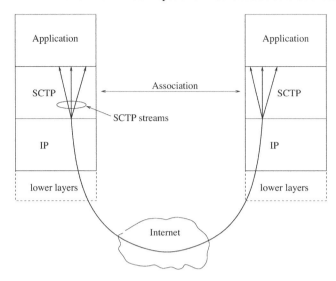

Figure 1.1 SCTP architecture showing association and SCTP streams.

1.2.1   SCTP Associations

Whereas TCP established a connection between two endpoints, thereby providing a connection-oriented service that guarantees that all packets are delivered in an ordered fashion, SCTP is said to create an *association* between two endpoints. The SCTP association has a broader scope than a simple connection: it encompasses multiple IP addresses to which a single SCTP port can be attached (as opposed to a single IP address to which a TCP port is bound). This allows an SCTP user to potentially receive and send data through all the available network interfaces without the need to juggle among several independent TCP connection. An association is established through a four-way handshake procedure between the two endpoints. It uses a cookie mechanism (Stewart 2007) to dispell DOS attacks.

1.2.2   SCTP Streams and Message Ordering

In TCP, data are exchanged as a single stream, i.e., the application perceives the service offered by TCP as that of a "pipe" where every bit entering one end is orderly extracted at the other end, without two or more bits ever being swapped. While this service is highly desirable in most cases, applications in need of exchanging data of different nature in a parallel fashion may run into HoL blocking when one piece of data goes missing and TCP strives to recover it through retransmission. In such a case, applications have no option other than setting up multiple independent TCP connections.

SCTP allows an application to set up multiple logical "streams" within an association and to specify which stream the data belongs to upon passing it to SCTP.

When data from one stream is missing at the receiver, the messages belonging to other streams are not held up for reordering, but are delivered to the application provided the ordering within those streams is maintained. The number of streams in an association is set by the four-way handshake that takes place during the association establishment.

As an additional option, an SCTP endpoint can indicate that no ordered delivery is necessary for a particular chunk within a stream by setting a flag in the chunk header. Upon reception, the chunk is immediately delivered to the upper layer, whether it has been received in order or not.

1.2.3 The Message-Oriented Nature of SCTP

As pointed out above, TCP is a stream-oriented protocol, and the data the endpoint exchanges are the result of several asynchronous `read` operations from an upper-layer stream (at the sender) and `write` operations into a similar stream handed over to the upper-layer (at the receiver). This forces applications using TCP to delineate message boundaries by using markings of sorts and by setting the PUSH flag in TCP segments. Conversely, SCTP is a *message-oriented* protocol: thus, all `read` primitives from the application layer return a single message to the SCTP layer, which in turn handles it through a single *send* operation to the other SCTP endpoint. There, a `write` primitive delivers the whole received message to the application layer. As a result, applications using SCTP are entitled to expect delivery of their data on a message-by-message basis.

1.2.4 Partial Reliability

One of the tenets of the TCP service model is the reliable delivery, which complements the utter lack of reliability offered by the IP protocol. However, as already remarked, this may limit the applicability of TCP if applications are not as concerned with missing a bunch of bits as they are with expediting the reception of the bulk of the data. This may have marginal relevance, or even be detrimental, for data traffic, but becomes fundamental for multimedia audio/video applications, where timely delivery is more important than full reliability. Though initially designed to offer a similar reliable service, SCTP saw a later addition to the standard, the Partial Reliability extension (SCTP-PR) (Stewart et al. 2004), that allows applications with stringent timing requirements to forego reliability in favor of timeliness. The PR extension can be activated by applications imposing an upper limit to the number of retransmission of a single SCTP data chunk after which the chunk is *advanced acknowledged*, that is to say, artificially acknowledged at the source without any actual ACK message reception. Thus, the SCTP sender window is allowed to keep moving, enabling further transmissions.

1.3   Multihoming and Its Support in SCTP

The term "multihoming" was not coined for SCTP, but has been around the Internet
for a while. Originally, it was used to refer to a specific type of Autonomous System
(AS), namely multihomed AS, which is connected to more than one AS but does
not allow packets from outside sources to transit on their way to an outside destination. The above definition relates to the availability of multiple links, or paths, that
can be chosen by a single packet while it is routed between source and destination.
This definition, however, does not make any assumptions on the number of network
interfaces that either the source or the destination have. This is not the type of multihoming that we are going to investigate in this chapter, rather we are interested in
multihoming involving hosts with multiple interfaces, each nominally attached to a
different network.

As stated in RFC 4960, "An SCTP endpoint is considered multihomed if there
are more than one transport address that can be used as a destination address to reach
that endpoint." In other words, SCTP stations with multiple network interfaces, each
identified by a separate IP address, can establish a single association between them
and use multihoming for redundancy purposes. Every station chooses a *primary* destination address, normally used for the transmission of new messages, whereas the
alternate addresses are considered as *secondary*, or backup, paths, whose conditions
are periodically monitored through the transmission of probe packets called *Heartbeats*.

This behavior is in stark contrast with what TCP does. Upon establishment, a
TCP connection is bound to a specific IP address. In case the IP address becomes
unreachable due to failure or mobility, the TCP connection stalls and is evantually
torn down. If a new IP address is assigned to the interface, the TCP connection must
thus be restarted. As such, TCP does not support multihoming, and whatever changes
occur to the point of attachment of a TCP host must be handled through techniques
that let the host keep its IP address, possibly through tunnelling or address mas-
querading. Few attempts at patching TCP to introduce multihoming support have
been made, but the topic never received wide attention from the networking research
community. TCP multihoming support will be further discussed in Section 1.4.

However, as will be clarified in the following, SCTP multihoming goes well beyond the mere support of IP address changes. Having introduced such flexibility
above the IP layer paved the way to a number of advanced features, such as the ability to select the best performing links across a range of different points of attachment,
or data striping through parallel transmissions on multiple paths.

1.3.1   Robustness to Path Failover

As already remarked, fault tolerance was the original purpose of SCTP multihoming,
long before concurrent transmission was considered. Introduced as early as in RFC
2960, it relies on the definition of primary and secondary (or backup) paths.

In SCTP, the backup paths are used only (i) to retransmit lost chunks, in order to increase the probability of successful retransmissions; (ii) to transmit new

chunks when, due to the expiration of several consecutive timeouts on the primary path, the primary interface is declared as "inactive." The retransmission counter (or Path.Max.Retrans, PMR) is maintained for each IP address at the destination and it allows SCTP to distinguish a path failure from temporary congestion of one of its links. In the case of detected failure, SCTP transmits new chunks toward a backup destination address and Heartbeat packets toward the primary one. As soon as the Heartbeats reception on the primary interface is confirmed, its state is toggled to "active" and the transmission over the primary path is resumed.

The setting of the PMR (RFC4960 recommends its value to be 5) is obviously a critical factor for failover performance: set it too high and path failure detection is delayed; if too small value is chosen, unnecessary failovers ensue. Several works have investigated its impact, either in the context of SS7 signalling (Grinnemo and Brunstrom 2004, 2005; Jungmaier et al. 2002), or within IP networks (Caro et al. 2004, 2006; Eklund et al. 2008). Findings pointed at the need to either lower the value of the PMR (maintaining that the impact of spurious failovers was not as dire as delayed failure detection) or to decrease the SACK timer of SCTP chunks. Additionally, Noonan et al. 2006 warns that the loss of SACKs may confuse an SCTP endpoint as to which path is operational and which has failed.

1.3.2 Optimal Path Selection

Beside exploiting it upon path failure, multihoming can come in handy to estimate which path should be selected as primary because the performance there (in terms of throughput, latency, loss rate) is superior. Such optimal path selection can, of course, occur in a dynamic fashion through constant probing and monitoring of primary and secondary paths, and through the introduction of appropriate metrics that facilitate the choice. Optimal path selection is instrumental to the definition of the so-called "Always Best Connected" (ABC) scenario, where multi-interface mobile nodes are allowed to take advantage of the simultaneous use of diverse access technologies.

It is to be remarked that, while traffic on the primary path is easily monitored, secondary paths only see sporadic transmissions of heartbeat packets, hence metrics collected there tend to be less reliable. Primary/secondary path swapping can thus occur several times during the lifetime of an association, triggered by one or more metrics collected by the sender on its own, or upon notification by the receiver. Optimal path selection features are especially desirable when one or more paths cross time-variant channels, i.e., when some links are of wireless nature. A typical setting in which optimal path selection is used can be seen in Figure 1.2.

An interesting approach to optimal path selection can be found in Kashihara et al. 2004, which suggests the use of packet-pair bandwidth estimation and Round Trip Time (RTT) measurements to trigger the selection. In Ma et al. 2004 multihoming was applied within a UMTS/WLAN overlay architecture to improve throughput performance, while Noonan et al. 2004 used delay and jitter over all paths, monitored through data traffic and heartbeats. In Li et al. 2011, optimal path selection in the case of competing sources was investigate and fair allocation of the paths capacity to them was proposed, based on network utility maximization.

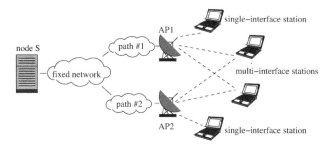

Figure 1.2 Optimal path selection scenario.

Optimal path selection has also been explored in the context of *autonomic* networking, spurring a number of works that have focused on SCTP multihoming, such as Casetti et al. 2008; Fracchia et al. 2007; Kamphenkel et al. 2009; Nguyen and Bonnet 2008. In Fracchia et al. 2007, a modification of the SCTP congestion control mechanism was introduced to discriminates losses due to congestion from losses due to channel conditions and to react accordingly by properly reducing the generated traffic only when needed. As a by-product, a dynamic redefinition of the identity of primary and secondary paths was introduced to allow transmissions on the path with the largest available bandwidth. Casetti et al. 2008 presents an autonomic solution to select which wireless interface (and, correspondingly, which network point of attachment) to be used for data transfer while a multihomed node is on the move by constantly monitoring the available bandwidth and the capacity between transport-layer endpoints over both primary and secondary paths. Machine-learning techniques were instead used in Kamphenkel et al. 2009 when the actual capacity of paths is not known in advance and no bandwidth estimation techniques can be employed, while (Nguyen and Bonnet 2008) provides a framework for intelligent tunneling using SCTP over multi-interface Mobile routers, in an ABC scenario.

1.3.3   Mobility Support

The Dynamic Address Reconfiguration (DAR) SCTP extension (Stewart et al. 2007) was introduced to support the dynamic addition and subtraction of IP addresses to an SCTP association without restarting it. Although, in the intentions of IETF, it was meant to cope with IPv6 renumbering, it quickly became clear that the DAR extension could also be used as a mobility-enabling feature that allowed an SCTP host to roam freely across IP subnets. Many researchers were quick to catch on the DAR extension as if it came to define an entirely new protocol, which is often referred to as mobile SCTP, or mSCTP (Chang et al. 2007; Koh et al. 2004a,b; Riegel et al. 2002). Not so the IETF, which, at the time of this writing, has not standardized it and has rejected several Internet Drafts on the topic.

Beside representing a new protocol, mSCTP had the merit of ushering in the new concept of *transport-layer* mobility (Atiquzzaman and Reaz 2005), as opposed to the established *network-layer* mobility. By pushing the mobility management to the end nodes, rather than burdening the core network with it, unhindered terminal mobility could become a reality (Eddy 2004). Indeed, unlike SIP-based handover schemes, mSCTP manages the handover at the transport layer without support from the routers. Quite simply, it dynamically adds or deletes IP addresses to and from end points engaged in the same association, and it replaces the primary path used by the association.

The work in Budzisz et al. 2008 discussed key scenarios and challenging issues in choosing mSCTP to handle seamless mobility at the transport layer in heterogeneous wireless access networks. Additionally, they showed the unsuitability of the legacy SCTP failover mechanism to handle mobility, especially for real-time services.

The possibility of using SCTP multihoming to reduce the network load after a Mobile IP handover has also been evaluated, for example in Zhang et al. 2007, where the home address and the care-of address are assigned to the mobile host. The home address becomes permanently attached to the SCTP association, while the care-of address is assigned by the current point of attachment to the network. The typical ill-effects of triangle routing in Mobile IP are removed by using the two addresses for the dual purpose of locating the host (home address) and exchanging data (care-of address).

1.3.4   Concurrent Multipath Transmissions and Data Striping

Unlike optimal path selection where an algorithm always tries to choose the best available path (according to one or more metrics) and funnels all traffic onto it, Concurrent Multipath Transmission (CMT) allows more than one path to carry packets from the same connection and with the same source-destination endpoints. However, appealing though it may sound, splitting a traffic flow over multiple connections was never a feature intended for SCTP when it was standardized. Indeed, most attempts at introducing modifications have run into several non-trivial problems related to message ordering.

First of all, architectural changes are required at the source: the SCTP sender must set up a buffer for each path, where messages are stored in preparation for the transmission on the scheduled path. The congestion control algorithm must then be applied to each of these buffer/path pair separately (a single congestion control would defeat the very purpose of having concurrent transmissions).

Secondly, the path diversity induced by different bandwidth and delay in each path in the association, may lead to a high number of *out-of-order* message arrivals at the receiver. If no changes are introduced in the congestion control mechanism, this causes the receiver to advertise large, lasting gaps in the received chunk sequence, flooding the network with immediate ACKs, forcing several fast retransmissions and triggering unneeded reductions of the congestion window. Scheduling algorithms that minimize the number of out-of-order messages, such as the one proposed in Fiore et al. 2007 and based on end-to-end bandwidth estimation, are called for.

Finally, out-of-order messages may rapidly fill small-sized receiver buffers, potentially leading to the stalling of the association (Iyengar et al. 2007b). A workaround is proposed by (Natarajan et al. 2009), suggesting that paths that experienced a loss are temporarily excluded from the message scheduling round until a heartbeat is received.

For the above reasons, researches into extending SCTP to CMT, currently in progress, agree on the need for structural changes to the original SCTP standard. In (Iyengar et al. 2006, 2004) several modifications at the sender are proposed to compensate for the problems introduced by using a unique sequence-number space for data transfers occurring concurrently over multiple paths. A multihomed sender maintains per-destination virtual queues and spreads chunks across all available paths as soon as the congestion window allows it. Retransmissions are triggered only when several selective acknowledgments (SACKs) report missing chunks from the same virtual queue.

Load-Sharing SCTP (LS-SCTP) (Al et al. 2004) aggregates the bandwidth of the transmission paths between the communicating endpoints, and dynamically adds new paths as they become available. Path monitoring is used to stripe packets among all available paths ensuring that the association does not stall as a consequence of high loss rates or temporary path unavailability. The key idea is to introduce a per-association, per-path data-unit sequence-numbering that extends the per-association SCTP congestion control to a finer-grained per-path congestion control. This, however, requires a change in the SCTP packet format, and modifications at both the sender and the receiver.

Independent Per-path Congestion Control SCTP (IPCC-SCTP) (Ye et al. 2004) builds upon the LS-SCTP idea of enhancing SCTP with a per-path congestion control, but tries to avoid modifications in the packet format and in the receiver by keeping more state information at the sender about which path the individual data units are sent on.

An interesting conclusion is reached in Iyengar et al. 2007a, where CMT and optimal path techniques are compared and the latter is shown to provide better performance in presence of largely asymmetric path lengths and different loss rates on each path.

The effects of multihoming and multipath on congestion control algorithms at the transport layer (focusing on TCP) have also been studied through the lenses of control theory in Han et al. 2006, where it was shown that stable congestion control can be fully extended from the single-path mode (as in TCP) to a multipath case using an overlay network of router or other multihoming techniques.

1.4   Multihoming and Multipath Support Outside SCTP

In spite of the large volume of research work on SCTP as a solution for multihoming and multipath, as reported in the previous Section, the deployment of SCTP on consumer devices is regrettably low. There are several reasons for it: wavering support by major OSs (SCTP is not bundled in either Windows, MAC OS, or in the mainstream Linux distributions), the lack of a properly designed session layer in the TCP/IP protocol stack which forces applications to state what protocol they need at the transport layer upon opening a socket (and thus requiring applications to be rebuilt for explicit SCTP support).

Such lack of interest is all the more conspicuous if one looks at the current coexistence of heterogeneous wireless networks and at the development, by many manufactures, of dual-mode or multimode handsets. One may naturally wonder if, given this trend, other technologies are under development, or are mature enough, to offer multihoming and multipath support. It is therefore interesting to take a sweeping look across all architectural layers and examine other proposals where SCTP is not involved.

1.4.1 Physical and Link Layer Solutions

Multihoming approaches first appeared with channel bonding schemes associating several physical links to enhance availability, reliability in the face of failover and increase overall throughput. Ethernet switches support channel bonding over several links as mandated by the IEEE P802.3ad standard. As a result, if a single link fails and several links are bundled together, no connection interruption is experience by the end appliactions. However, Ethernet bonding is a local solution and it mainly addresses wired networks.

Layer-two aggregation has also been considered for broadband access with Multilink PPP, standardized in RFC 1990 (Sklower et al. 1996). Multilink PPP handles several connections and coordinates multiple independent links between a fixed pair of systems, providing a virtual link with greater bandwidth than any of the constituent members. The aggregate link, or bundle, is transparent to upper layers and it can optimize the transmission by splitting packets among parallel virtual circuits. Such a solution was originally used to handle multiple ISDN circuits at basic and primary rate, and has found little use ever since.

1.4.2 Network Layer Solutions

When looking at the network layer, it is of course necessary to mention Mobile IP, defined in RFC 5944 (Perkins 2010) as the current IETF standard for supporting mobility in the Internet. Mobile IP introduces a level of indirection into the routing architecture of the IP protocol. The home address of a mobile host loses its function as an interface identifier and becomes an end-point identifier. Mobile IP creates a routing tunnel between a mobile host's *home address* and its *care-of address*. Thus, Mobile IP ensures the delivery of packets destined to the home address of a mobile host, regardless of its physical point of attachment to the Internet, which is reflected by its care-of address. For our current study, however, it is to be remarked that the two addresses cannot be used independently to support concurrent multipath transmissions. Several other architectures have been proposed for approaches relying on IP, though often they require the presence of proxies, as detailed in Bichot et al. 2005; Mao et al. 2005; Seth et al. 2005.

The IEFT has been at the forefront of multihoming activity even outside SCTP, notably through the Monami6 working group, which extended Mobile IP protocols to support multihoming, enabling the terminal to distribute multiple communications over several tunnels. However, they did not consider using several tunnels for a single data flow and they retained traditional centralization and tunneling approaches potentially leading to network bottlenecks or a single point of failure (Wakikawa et al. 2009). Other multihoming protocols, such as SHIM6 (Nordmark and Bagnulo 2009) and HIP (Moskowitz et al. 2008) allow the set-up of contexts between two end-hosts and the exchange of a set of addresses to be used for communication, though simultaneous use of these addresses is not envisaged.

1.4.3   Transport Layer Solutions

Multihoming solutions explicitly positioned at the transport layer were mainly concerned with preventing TCP from dropping the connection upon lower-layer disconnections. As an example, in Snoeren and Balakrishnan 2000, an end-to-end approach called *TCP migrate* uses a special Migrate SYN packet that contains the token identifying a previous connection after the host has moved to a different IP address. A previously-established TCP connection is thus restarted from the new address without losing its state (sequence numbers, ACKs and retransmission counts).

Multipath TCP has recently attracted novel interest in the field of wireless mesh networks where solutions compounding TCP over multiple mesh paths and network coding were explored (Gheorghiu et al. 2009; Xia et al. 2009).

The obvious limitations of such solutions (that have not been implemented on a large scale, nor accepted as a standard) is that they are only TCP-specific and cannot be extended to applications not running on top of TCP.

1.4.4   Higher Layer Solutions

In what is likely to be the most successful attempt at supporting multihoming in consumer devices, IEEE has been developing the 802.21 standard (IEEE802.21 2009) and defining a media-independent entity that provides a generic interface between the different link layer technologies and the upper layers. IEEE 802.21 aims at facilitating seamless handover between heterogeneous access technologies (such as WLANs, cellular or wireline networks) and is designed to work with different mobility management mechanisms. To handle the particularities of each technology, 802.21 maps its generic interface to a set of media-dependent service access points (SAPs) whose aim is to collect information and control link behavior during handovers. Additionally, the IETF has chartered a working group to address media-independent handover (MIH) in IP networks (mipshop charter 2008) that is designing a layer-3 protocol for transporting MIH-related information along with DHCP and DNS extensions for discovering the information servers within the 802.21 framework. While the new 802.21 standard is specifically designed with seamless handover in mind, it is also being envisaged to support and facilitate the implementation of traditional concurrent multipath solutions at the transport layer, as in Fallon et al. 2010; Sarkar and

Sarkar 2009.

1.5 Closing Remarks

In the face of the growing number of access network technologies and of multi-mode consumer communication devices, the networking community seems to be in a quandary as to how to provide a definitive answer to the problem of multihomed terminals and multipath communication. In this chapter we have outlined the various flavors of what is the natural candidate to provide such answer, the SCTP protocol. We have shown how by design and purpose, SCTP could single-handedly provide end-to-end multihoming and multipath support, to the point of being a potential replacement of the TCP protocol.

At the same time, we have tried to identify the many challenges and hurdles that still hamper the "rise to stardom" of SCTP. Indeed, SCTP is a protocol of many potentialities and unfulfilled promises: despite being an established IETF standard, and thus fully integrated in the Internet protocol stack, it is all too often relegated as an optional feature in real-life networking implementations, which severely limits its widespread diffusion.

Bibliography

Al AA, Saadawi T and Lee M 2004 LS-SCTP: a bandwidth aggregation technique for stream control transmission protocol. *Computer Communications* **27**(10), 1012–1024.

Atiquzzaman M and Reaz A 2005 Survey and classification of transport layer mobility management schemes. *16th IEEE International Symposium on Personal Indoor and Mobile Radio Communications (PIMRC)*.

Bichot G, Brauneis W and Linder H 2005 New architectures for an enhanced multimedia wireless broadband service. *Broadband Europe*.

Budzisz L, Ferrús R, Brunstrom A, Grinnemo KJ, Fracchia R, Galante G and Casadevall F 2008 Towards transport-layer mobility: Evolution of SCTP multihoming. *Computer Communications* **31**(5), 980–998.

Caro A, Amer P and Stewart R 2004 End-to-end failover thresholds for transport layer multihoming. *IEEE MILCOM*.

Caro A, Amer P and Stewart R 2006 Rethinking end-to-end failover with transport layer multihoming. *Annales des Telecommunications/Annals of Telecommunications* **61**(1), 92–114.

Casetti C, Chiasserini CF, Fracchia R and Meo M 2008 Autonomic interface selection for mobile wireless users. *IEEE Transactions on Vehicular Technology* **57**(6), 3666–3678.

Chang LH, Lin HJ and Chang I 2007 Dynamic handover mechanism using mobile SCTP in contention based wireless network. *LNCS 4742* pp. 821–831.

Eddy M 2004 At what layer does mobility belong? *IEEE Communications Magazine* **42**(10), 155–159.

Eklund J, Brunstrom A and Grinnemo KJ 2008 On the relation between sack delay and SCTP failover performance for different traffic distributions. *The 5th International Conference on Broadband Communications and Networks and and Systems (BROADNETS)*.

Fallon E, Qiao Y, Murphy L, Murphy J and Muntean G 2010 SOLTA: a service oriented link triggering algorithm for mih implementations. *ACM IWCMC*.

Fiore M, Casetti C and Galante G 2007 Concurrent multipath communication for real time traffic. *Computer Communications* **30**(17), 3307–3320.

Fracchia R, Casetti C, Chiasserini CF and Meo M 2007 Wise: Best-path selection in wireless multihoming environments. *IEEE Transactions on Mobile Computing* **6**(10), 1130–1141.

Gheorghiu S, Toledo AL and Rodriguez P 2009 Multipath TCP with network coding for wireless mesh networks. *IEEE ICC*.

Grinnemo KJ and Brunstrom A 2004 Performance of SCTP-controlled failovers in M3UA-based SIGTRAN networks. *Advanced Simulation Technologies Conference (ASTC)*.

Grinnemo KJ and Brunstrom A 2005 Impact of traffic load on SCTP failovers in sigtran. *International Conference on Networking (ICN)*.

Han H, Shakkottai S, Hollot CV, Srikant R and Towsley D 2006 Multi-path TCP: A joint congestion control and routing scheme to exploit path diversity in the internet. *IEEE/ACM Trans. on Networking* **14**(6), 1260–1271.

IEEE802.21 2009 http://www.ieee802.org/21/.

Iyengar J, Amer P and Stewart R 2006 Concurrent multipath transfer using SCTP multihoming over independent end-to-end paths. *IEEE/ACM Transactions on Networking* **14**(10), 951–964.

Iyengar JR, Amer PD and Stewart R 2007a Performance implications of a bounded receive buffer in concurrent multipath transfer. *Computer Communications* **30**(4), 818–829.

Iyengar JR, Amer PD and Stewart R 2007b Performance implications of receive buffer blocking in concurrent multipath transfer. *Computer Communications* **30**(4), 818–829.

Iyengar JR, Shah KC, Amer PD and Stewart R 2004 Concurrent multipath transfer using SCTP multihoming. *SPECTS*.

Jungmaier A, Rathgeb E and Tuexen M 2002 On the use of SCTP in failover scenarios. *6th World Multiconference on Systemics and Cybernetics and and Informatics (SCI)*.

Kamphenkel K, Laumann S, Bauer J and Carle G 2009 Path selection techniques for SCTP multihoming. *IEEE ICC Workshops*.

Kashihara S, Nishiyama T and Iida K 2004 Path selection using active measurement in multi-homed wireless networks. *IEEE SAINT*.

Koh SJ, Chang MJ, J. M and Lee M 2004a mSCTP for soft handover in transport layer. *IEEE Communications Letters* **8**(3), 189–191.

Koh SJ, Jung HY and Min JH 2004b Transport layer internet mobility based on mSCTP. *IEEE International Conference on Advanced Communication Technology*.

Li S, Qin Y and Zhang H 2011 Distributed rate allocation for flows in best path transfer using SCTP multihoming. *Telecommunication Systems* **46**(1), 81–94.

Ma L, Yu F and Leung VCM 2004 A new method to support UMTS/WLAN vertical handover using SCTP. *IEEE Wireless Communications*.

Mao Y, Knutsson B, Lu H, and Smith J 2005 Dharma: Distributed home agent for robust mobile access. *IEEE INFOCOM*.

mipshop charter 2008 http://www.ietf.org/html.charters/mipshop-charter.html.

Moskowitz R, Nikander P, Jokela P and Henderson T 2008 RFC 5201: Host identity protocol. *IETF*.

Natarajan P, Ekiz N, Amer PD and Stewart R 2009 Concurrent multipath transfer during path failure. *Computer Communications* **32**(15), 1577–1587.

Nguyen H and Bonnet C 2008 An intelligent tunneling framework for always best connected support in network mobility (nemo). *IEEE WCNC*.

Noonan J, Perry P, Murphy S and Murphy J 2004 Client controlled network selection. *5th IEE International Conference on 3G Mobile Communication Technologies*.

Noonan J, Perry P, Murphy S and Murphy J 2006 Stall and path monitoring issues in SCTP. *IEEE Infocom*.

Nordmark E and Bagnulo M 2009 RFC 5533: Shim6: Level 3 multihoming shim protocol for ipv6. *IETF*.

Perkins C 2010 RFC 5944: Ip mobility support for IPv4. *IETF*.

Riegel M, Tuxen M, Rozic N and Begusic D 2002 Mobile SCTP transport layer mobility management for the internet. *SoftCOM*.

Sarkar D and Sarkar U 2009 Balancing load of aps by concurrent association of every wireless node with many aps. *IEEE ICNS*.

Seth A, Darragh P, Liang S, Y.Lin, and Keshav S 2005 An architecture for tetherless communication. *Proc. of Dagstuhl DTN Workshop*.

Sigtran 2007 http://toolsietf.org/wg/sigtran/.

Sklower K *et al.* 1996 RFC 1990: The ppp multilink protocol. *IETF*.

Snoeren A and Balakrishnan H 2000 An end-to-end approach to host mobility. *ACM MOBI-COM*.

Stewart R 2007 RFC 4960: Stream control transport protocol. *IETF*.

Stewart R *et al.* 2004 RFC 3758: Stream control transport protocol (SCTP) - partial reliability extension. *IETF*.

Stewart R, Xie Q, Tuexen M, Maruyama S and M.Kozuka 2007 RFC 5061: Stream control transmission protocol (SCTP) dynamic address. *IETF*.

Wakikawa R *et al.* 2009 RFC 5648: Multiple care-of addresses registration. *IETF*.

Xia Z, Chen Z, Ming Z and Liu J 2009 A multipath TCP based on network coding in wireless mesh networks. *IEEE ICISE*.

Ye G, Saadawi TN and Lee M 2004 IPCC-SCTP: an enhancement to the standard SCTP to support multi-homing efficiently. *IEEE International Conference on Performance and Computing and and Communications (ICPCC)*.

Zhang J, Chan H and Leung V 2007 A sip-based seamless-handoff (s-sip) scheme for heterogeneous mobile networks. *IEEE Wireless Communications and Networking Conference (WCNC)* pp. 3946–3950.

2

Fault Tolerance

Armando Caro

This chapter focuses on transport layer techniques that exploit host multihoming to provide end-to-end fault tolerance and improved application performance. Generally, mission critical systems rely on redundancy at multiple levels to provide uninter-

rupted service during resource failures. Following the redundancy approach, mission
critical systems can multihome their hosts to improve availability. A host is multihomed if it can be addressed by multiple IP addresses (Braden 1989b). Redundancy
at the network layer allows a host to be accessible even if one of its IP addresses
becomes unreachable for an extended period of time (assuming the paths to the multiple interfaces do not share the same failed link). Transport layers that support multihoming allow traffic of existing connections to be redirected to a peer's alternate IP
address without the need for applications (or users) to abort and re-establish connections. Considering the prevalence of path outages on the Internet today, multihoming
support at the transport layer can improve resilience of established connections, and
thus improve application performance. While fault tolerance can be addressed at
other layers, we argue that the transport layer is in the best position to detect failure
and make end-to-end failover decisions. After all, the transport layer is the lowest
layer responsible for both end-to-end quality of service and having knowledge about
end-to-end path characteristics.

2.1   Introduction

Widespread use of multihoming was infeasible during the early days of the Internet due to cost constraints; today, network interfaces have become commodity items.
Cheaper network interfaces and cheaper Internet access motivate content providers to
have simultaneous connectivity through multiple Internet Service Providers (ISPs),
and more home users are installing wired and wireless connections for added flexibility and fault tolerance. Furthermore, wireless devices are being simultaneously connected through multiple access technologies, such as wireless LANs (e.g., 802.11)
and cellular networks (e.g., GPRS, CDMA).

2.1.1   Motivation for Multihoming

We begin with an argument for why host multihoming is important today and is expected to further increase in importance in the short and long terms. Today, content
providers and server operators on the Internet face a significant challenge of ensuring
that services are readily available and highly reliable. Servers on the Internet today
are recognized as being much less reliable than mission-critical services provided
today, such as the telephone network's "five 9's" expected level of reliable operation.
Much of the gap in reliability is attributed to network availability and host reachability; unfortunately, network path outages are not rare on the Internet. We present two
causes of path outages on the Internet: link failures and overloaded links.

Link Failures

The Internet's backbone routing is designed to be robust against link failures. Link
failures occur when a router or a link connecting two routers fails due to link disconnection, hardware malfunction, or software error. The routing system has the

responsibility to detect such failures, and in response, reconfigure routing tables to bypass the failure.

One problem with routing recovery is that the Internet's backbone routing, which is based on Border Gateway Protocol (BGP) (Rekhter and Li 1995; Rekhter et al. 2002), has been optimized for scalability and simplicity. As a result, BGP can take a long time to converge on a new route after a link failure is detected. Labovitz et al. (2000) show that the Internet's interdomain routers may take tens of minutes to converge after a failure. They find that during these "delayed convergences," end-to-end internet paths experience intermittent loss of connectivity in addition to increased packet loss, latency, and reordering. Although they demonstrate that much of the convergence delays are due to BGP specification ambiguity and specific router vendor implementation decisions, long convergence times are inherent to the path vector routing protocol. Their analysis shows that in the best case, convergence times will grow linearly with the addition of new autonomous systems; in the worst case, the growth is exponential.

Paxson (1997) uses probes to find that "significant routing pathologies" prevent selected pairs of hosts from communicating about 1.5% to 3.3% of the time. Importantly, he also finds that this trend has not improved with time. Labovitz et al. (1999) examine routing table logs of internet backbones to find that 10% of all considered routes were available less than 95% of the time, and more than 65% of all routes were available less than 99.99% of the time. They also find that the duration of path outages are heavy-tailed, and about 40% of path outages take more than 30 minutes to repair. Chandra et al. (2001) use probes to confirm that failure durations are heavy-tailed, and report that 5% of detected failures last more than 2.75 hours, and as long as 27.75 hours.

Overloaded Links

Another problem with BGP is that it is not designed to detect performance failures due to overloaded network links. Flash crowds and Denial-of-Service (DoS) attacks are two situations that can overload a path to a host. Such scenarios drastically degrade the end-to-end communication between peer hosts, but BGP is unaware of such performance failures. Even if alternate paths exist, BGP will not use them to route traffic around overloaded links.

A flash crowd is a sudden, large surge of legitimate traffic to a particular site. Flash crowds resulted from the Ken Starr Report on September 11, 1998, Victoria's Secret Fashion Show webcast on May 18, 2000, and news reports of the terrorist attacks of September 11, 2001. Also, the *Slashdot Effect* is a commonly experienced phenomenon that occurs when a high-volume news web site (www.slashdot.org) posts a web site link and triggers a spontaneous high hit rate upon the server of that link.

A DoS attack floods a network with useless traffic to overwhelm the network and/or system resources of the victim host. Moore et al. (2001) study three, one-week datasets to find that DoS activity on the Internet is widespread, and distributed among many different domains and ISPs. During the three one-week periods, they observe more than 12,800 attacks against more than 5,000 distinct targets belonging to 2,000 distinct DNS domains. They report attack rates and durations that would disrupt most (if not all) communication to the victim IP addresses. They report that 50% of attacks are less than 10 minutes in duration, 80% are less than 30 minutes, and 90% less than an hour. At the tail of the distribution, 2% of attacks are greater than 5 hours, 1% are greater than 10 hours, and dozens last multiple days. They also find that the range of targets is surprisingly more encompassing than expected. Targets include well known sites, such as Amazon and Hotmail, but a significant fraction of attacks is directed against small and medium sized businesses, dialup and broadband home machines, and network infrastructure, such as name servers and routers.

These statistics are not promising for mission critical systems which require high availability. Many commercial services advertise 99.99% or 99.999% (known as "four 9's" or "five 9's) server availability, but highly available servers do not guarantee highly available services for their clients. The end-to-end paths from the clients to the server must also be highly available. A system which advertises 99.999% server availability expects to have a down time of at most 5 minutes a year. However, a typical client may find the server available much less often; even 99% availability translates to approximately 15 minutes of down time a day.

Previous research has attempted to improve host availability on the Internet by various means, such as server replication (Akamai 1998; Crovella and Carter 1995; Dykes et al. 2000; Freedman et al. 2004; Tuexen et al. 2002; Yoshikawa et al. 1997), site multihoming (Abley et al. 2003; Avaya: Adaptive Networking Software (Route-Science) 2004; Fat Pipe Networks 1989; Internap: Network-based Route Optimization 1996; Radware 1997; Stonesoft: StoneGate Multi-link Technology 1990), or overlay routing networks (Andersen 2005; Andersen et al. 2001; Gummadi et al. 2004; Yip 2002). Each of these approaches can be effective, but they have limitations (see Section 2.4.1). Server replication's costly infrastructure limits this solution to high-end web sites that can afford the expense. Site multihoming (provisioning a site with multiple ISP links) protects against a single access link failure, but it cannot avoid the long convergence times of BGP within the Internet. Overlay routing networks require extra infrastructure within the network that does not scale well beyond a small set of nodes. Furthermore, each of these approaches are unable to route around last-hop failures, which Gummadi et al. (2004) show to occur in 16% of path failures to servers and 60% of path failures to broadband hosts. Host multihoming, on the other hand, can route around last-hop failures by allowing a host to be accessible even if one of its IP addresses becomes unreachable.

2.1.2   Transport Layer Multihoming

Transport layer multihoming is a feature that binds a single transport layer connection to multiple network addresses at each endpoint. This transport protocol feature is not a new concept; it is actually an old concept disguised under a new name. Historically, transport layer multihoming was referred to as splitting/recombining or downward-multiplexing, and was for providing added resilience against network failure and/or potentially increasing throughput (Iren et al. 1999; Strayer and Weaver

1988; Walrand 1991). Although transport layer multihoming is an old concept, neither of the Internet's current transport protocol workhorses, TCP or UDP, support multihoming. UDP's connectionless nature is incompatible with transport layer multihoming, and TCP allows applications to bind to only one network address at each end of a connection. Furthermore, we are unaware of any other historical transport protocols that support multihoming (Iren et al. 1999). Network interfaces were expensive components in the early days of the Internet, which meant that transport layer multihoming was beyond the ken of research.

Now that network interfaces have become commodity items, and multiple ISP access (e.g., dialup/broadband home connection, WiFi hotspots, and cellular provider) is affordable, transport layer multihoming has become a feature worth supporting and investigating. Two recent IETF transport layer protocols, the Stream Control Transmission Protocol (SCTP) (Caro et al. 2003; Stewart and Xie 2001; Stewart et al. 2000) and the Datagram Congestion Control Protocol (DCCP) (Kohler et al. 2004) support multihoming at the transport layer. DCCP is currently in its formative stages, and transport layer mobility is its driving motivation for multihoming support. SCTP, on the other hand, is more mature (RFC2960 was published in October 2000), and supports multihoming primarily for network fault tolerance. Therefore, we use SCTP for our experiments, but we believe the results and conclusions presented in this chapter apply in general to reliable transport protocols that support multihoming.

SCTP was originally developed to carry telephony signaling messages over IP networks. With continued work, SCTP evolved into a general purpose transport protocol that includes advanced delivery options. SCTP is a message-oriented protocol that provides a reliable, full-duplex connection, called an *association*. Its key features are multihoming, multistreaming, and extra protection against SYN-flooding attacks, blind masquerade attacks, and stale SCTP packets from previous associations. *Multistreaming* allows for independent delivery among streams, and reduces the risk of head-of-line blocking. SCTP resists SYN-flooding attacks by using a four-way handshake that verifies the legitimacy of an association initialization request before allocating server-side resources. Blind masquerade attacks and stale packets are avoided by using a 32-bit verification tag to validate the sender of an SCTP packet. Refer to Table 2.1 for a complete list of SCTP's services/features, compared and contrasted with TCP and UDP.

To explain SCTP's multihoming feature, we contrast SCTP and TCP in a multihomed topology (see Figure 2.1). Since TCP can only bind to a single IP address at each endpoint, four distinct TCP connections are possible between Hosts A and B: (A_1, B_1), (A_1, B_2), (A_2, B_1), (A_2, B_2). SCTP's binding, on the other hand, is not limited to a single IP address on each end. Instead, a single SCTP association may consist of all IP addresses, which in our example would be: $(\{A_1, A_2\}, \{B_1, B_2\})$.[1] Currently, SCTP uses multihoming for redundancy purposes only and not for concurrent multipath transfers (Iyengar et al. 2004a,b,c). Each endpoint chooses a single destination address as the primary destination address, which is used for all data traffic during "normal transmission." Note that a single port number is used at each

---

[1] Although less useful for network failure recovery, an SCTP endpoint may bind to a proper subset of its available IP addresses.

Table 2.1 Compare and contrast SCTP, TCP, and UDP

Services / Features	SCTP	TCP	UDP
Connection-oriented	yes	yes	no
Full duplex	yes	yes	yes
Reliable data transfer	yes	yes	no
Partial-reliable data transfer	optional	no	no
Flow control	yes	yes	no
TCP-friendly congestion control	yes	yes	no
ECN capable	yes	yes	no
Ordered data delivery	yes	yes	no
Unordered data delivery	yes	no	yes
Uses selective ACKs	yes	optional	no
Preservation of Application PDU Boundaries	yes	no	yes
Application PDU fragmentation	yes	yes	no
Application PDU bundling	yes	yes	no
Path MTU discovery	yes	yes	no
Multistreaming	yes	no	no
Multihoming	yes	no	no
Protection against SYN flooding attack	yes	no	n/a
Allows half-closed connections	no	yes	n/a
Reachability check	yes	yes	no
Pseudo-header for checksum	no (uses vtags)	yes	yes
Time wait state	for vtags	for 4-tuple	n/a

endpoint regardless of the number of IP addresses.

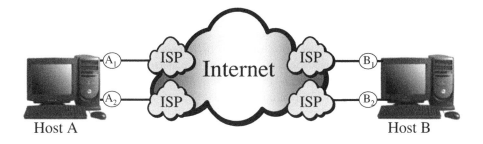

Figure 2.1 Example multihomed topology.

Now suppose that there exists a TCP connection, (A_1, B_1), and an SCTP association, $(\{A_1, A_2\}, \{B_1, B_2\})$, with A_1 and B_1 as the primary destinations. If B_1 becomes unreachable, TCP and SCTP behave quite differently. Since TCP does not support multihoming, the TCP connection awaits a predefined maximum number of retransmission timeouts and then aborts; thus forcing the application layer (or user) to recover. This "wait, abort, and reconnect" behavior may be unacceptable for mission critical applications. An SCTP association, however, remains alive even when B_1 becomes unreachable. SCTP's built-in failure detection and recovery system, known as *failover*, allows endpoints to dynamically send traffic to an alternate peer IP address when needed. Hence, in this example, the SCTP association temporarily redirects traffic to B_2 until B_1 becomes reachable again.

2.1.3 Chapter Overview

Transport layer multihoming is an old concept finally making its way into standardized transport protocols, but it remains a feature left mostly unexplored. With multiple paths at its disposal, a sending transport layer has several actions that can be taken to handle normal data transmission, loss recovery, path failure detection, failover, and path recovery. The authors of SCTP designed the fault tolerance mechanisms based on intuition and previous experience in single homed networking, but much of the multihoming functionality does not have research data to justify its design. This chapter investigates and challenges some of the design decisions of SCTP to determine if their predicted benefits are indeed realized. This research provides insight for future transport protocols that support multihoming.

Without complete knowledge about the network, a sender is faced with the dilemma of blindly deciding which path to use for loss recovery. A sender has no way of knowing if lost data are attributed to noise, minor transient congestion, major long term congestion, or path failure. Furthermore, endpoints are generally ignorant of the bandwidth-delay product of the available paths to their peer. Instead of gambling on a case-by-case basis, SCTP uses a fixed retransmission policy that attempts to maximize long term benefits. We challenge SCTP's retransmission policy by investigating a few policies for failure and non-failure scenarios. The results of this study are available in Section 2.2.

Path failure detection time is important for swift failovers that minimize stall time of transfers. Failure detection time can be improved by lowering the failover threshold, but doing so increases the number of false alarms (i.e., spurious failovers). We formally specify and evaluate a few variations of SCTP's failover mechanism to investigate the tradeoff between more aggressive failover and more frequent spurious failovers to determine appropriate failover thresholds for fast failovers that do not degrade goodput performance of transfers. The results of this study are available in Section 2.3.

We also challenge SCTP's failover mechanism for its temporary nature. As currently specified in RFC2960, failovers are never permanent; if the primary path recovers, a sender resumes sending new data to the primary destination. When changing destinations, a sender throttles its sending rate back to slow start to regain its ack clock for the new path. Consequently, the transfer is actually penalized for a path recovering. We formally specify and investigate extending the current state-of-the-art in multihoming to support *permanent failover* for avoiding such penalization. The results of this study are also available in Section 2.3.

Section 2.4 concludes this chapter. This section first presents a summary of related work that addresses end-to-end network fault tolerance. These approaches include dynamic host routing, site multihoming, server replication, overlay routing networks, and interrupted connection re-establishment. This sectin then presents a summary of key results in this chapter and suggestions for future study.

2.2   Retransmission Policies

Currently, an SCTP sender uses an alternate destination address for retransmissions when data sent to the primary destination is lost. SCTP's current retransmission policy (Stewart et al. 2000) states that "when its peer is multihomed, an endpoint SHOULD try to retransmit [data] to an active destination transport address that is different from the last destination address to which the [data] was sent." According to the authors of SCTP, this policy, which we refer to as *AllRtxAlt (All Retransmissions to Alternate)* attempts to improve the chance of success by sending retransmissions to alternate destinations (Stewart and Xie 2001). The underlying assumption is that loss indicates either that the network path to the primary destination is congested, or the primary destination is unreachable. Thus, retransmitting to an alternate destination will more likely avoid a repeated loss of the same data.

We show that this policy actually degrades performance in many circumstances. We explore two alternative retransmission policies that were introduced by this author and find that the best policy, for both failure and non-failure scenarios, is a hybrid policy: send (a) fast retransmissions to the primary destination, and (b) timeout retransmissions to an alternate destination. To this author's best knowledge, no transport protocol (necessarily multihomed) has ever explored such a hybrid policy. We show that this hybrid policy performs best when combined with two enhancements: a Multiple Fast Retransmit algorithm (also introduced by this author), and either timestamps or our Heartbeat After RTO mechanism (suggested by Randall Stewart). The Multiple Fast Retransmit algorithm reduces the number of timeouts. Timestamps and the Heartbeat After RTO mechanism both improve performance when timeouts are common by providing extra RTT measurements to increase a sender's accuracy of its RTT estimates and therefore RTO values.

Section 2.2.1 demonstrates a problem with SCTP's current retransmission policy (AllRtxAlt) by comparing it to an alternative policy, *AllRtxSame (All Retransmissions to Same)*. Section 2.2.2 introduces and evaluates our hybrid policy, *FrSameRtoAlt (Fast Retransmissions to Same, Timeouts to Alternate)*, which combines the good points of AllRtxAlt and AllRtxSame. Section 2.2.3 introduces and evaluates three extensions to further improve the performance of the three policies. Section 2.2.4 compares the policies' performance with their best extensions in non-failure scenarios, and Section 2.2.5 compares them in failure scenarios. Section 2.2.6 summarizes our retansmission policy results.

2.2.1   AllRtxAlt's Problem

AllRtxAlt is the retransmission policy currently specified for SCTP in RFC2960. This policy attempts to bypass transient network congestion and path failures by sending all retransmissions to an alternate destination. Intuitively, the author of this policy expected that sending retransmissions to an alternate path would be beneficial, particularly when the alternate path's quality is better (i.e., higher bandwidth, lower delay, and/or lower loss). Similarly, when the alternate path's quality is worse, one would expect sending retransmissions to the same destination as their original transmission should provide better performance. To test these hypotheses, we evaluate the performance of AllRtxAlt and the AllRtxSame policy – send all retransmissions to the same destination as their original transmission.

Analysis Methodology

We evaluate the retransmission policies using University of Delaware's SCTP module (Caro and Iyengar 2006) for the ns-2 network simulator (Berkeley et al. 2003). Figure 2.2 illustrates the network topology simulated: a dual-dumbbell topology whose core links have a bandwidth of 10Mbps and a one-way propagation delay of 25ms. Each router, *R*, is attached to five edge nodes. One of these five nodes is a dual-homed node for an SCTP endpoint, while the other four are single-homed and introduce cross-traffic that creates loss for the SCTP traffic.

The links to the dual-homed nodes have a bandwidth of 100Mbps and a one-way propagation delay of 10ms. The single-homed nodes also have 100Mbps links, but their propagation delays are randomly chosen from a uniform distribution between 5-20ms. The end-to-end one-way propagation delays range between 35-65ms. These delays roughly approximate reasonable Internet delays for distances such as coast-to-coast of the continental United States, and eastern United States to/from western Europe. Also, each link (both edge and core) has a buffer size twice the link's bandwidth-delay product, which is a reasonable setting in practice.

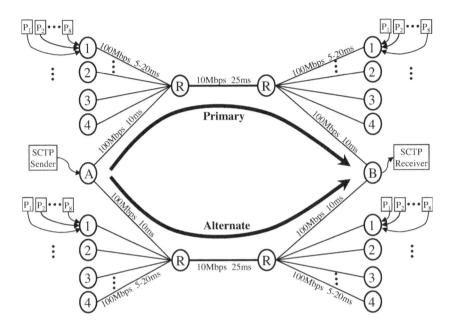

Figure 2.2 Simulation network topology with cross-traffic, congestion-based loss, and no failures.

Our configuration has two SCTP endpoints (sender *A*, receiver *B*) on either side of the network, which are attached to the dual-homed edge nodes. *A* has two paths, labeled primary and alternate, to *B*. Each single-homed edge node has eight traffic generators, each exhibiting ON/OFF patterns with ON-periods and OFF-periods drawn from a Pareto distribution. The cross-traffic packet sizes are chosen to *roughly* resemble the distribution found on the Internet: 50% are 44 bytes, 25% are 576 bytes, and 25% are 1500 bytes (CAIDA: Packet Sizes and Sequencing 1998; Claffy et al. 1998). The aim is to simulate an SCTP data transfer over a network with self-similar cross-traffic, which resembles the observed nature of traffic on data networks (Leland et al. 1993). We chose this model for simulating self-similar cross-traffic based on results by Willinger et al. (1995), which show that self-similar traffic can be modeled as an aggregation of ON/OFF sources with durations drawn from distributions with heavy tails (e.g., Pareto).

We simulate a 4MB file transfer with different network conditions, controlled by varying the load introduced by cross-traffic. All loss experienced is due to congestion at the routers; no loss is due to bit errors. The aggregate levels of cross-traffic on each path range from 5Mbps to 11Mbps. Although we independently control the levels of cross-traffic on each of the core links, the controls for the cross-traffic on each forward-return path pair are set the same. Each simulation has three parameters:

1. level of cross-traffic (in Mbps) on the primary path

2. level of cross-traffic (in Mbps) on the alternate path

3. AllRtxAlt vs AllRtxSame policy

Results

We compare the transfer times using AllRtxAlt versus AllRtxSame under various loss rates, with all else being equal (bandwidth, delay, etc.). Since loss in our simulations (only) occurs due to congestion, we do not set the loss rate. Instead, we set various levels of cross-traffic and calculate the observed loss rate for a transfer after the simulation has completed. The loss rate is calculated as the number of SCTP packets dropped by the routers (labelled R) divided by the number of SCTP packets transmitted by sender A.

We collected results for 0-10% loss on the primary and alternate paths, but for readability we do not include all results. Figure 2.3 presents the results for transfers with $\{3,5,8\}\%$ primary path loss. The graphs compare the file transfer time using AllRtxAlt versus AllRtxSame at various loss rates on the alternate path. Without failures, AllRtxSame never uses the alternate path, and therefore is unaffected by the alternate path's loss rate. Thus, AllRtxSame's transfer times are represented as a band parallel to the x-axis. This band outlines the upper and lower bounds of the 90% confidence interval. For example, we are 90% confident that the average 4MB file transfer time at 3% primary path loss lies between 34.3 and 35.1 seconds.

AllRtxAlt's transfer times are grouped by ranges of alternate path loss rates. The graph depicts the mean and the 90% confidence interval for each of these groups. The 90% confidence interval is calculated using an acceptable error of 10% of the mean. That is, we ran enough simulations to estimate the mean and 90% confidence interval with an acceptable error of at most 10% of the mean. For example, when the primary path's loss rate is 3% and the alternate path's loss rate is 1.5-2.5%, the 4MB file transfer time is on average 42.8 seconds with a 90% confidence interval between 41.1 and 44.5 seconds.

The graphs show that for $\{3,5\}\%$ primary path loss, AllRtxSame outperforms AllRtxAlt for all alternate path loss rates (except 0%). *Even at times when the alternate path's loss rate is better (i.e., lower) than the primary's, retransmitting on the alternate path degrades performance.* This trend remains for all loss rates. Consider the results for 8% primary path loss. The anticipated benefits of AllRtxAlt only appear for alternate path loss rates of 0-3%. In other words, even if the alternate path's loss rate is up to 3% better (5-8%), it is better to retransmit data on the primary path

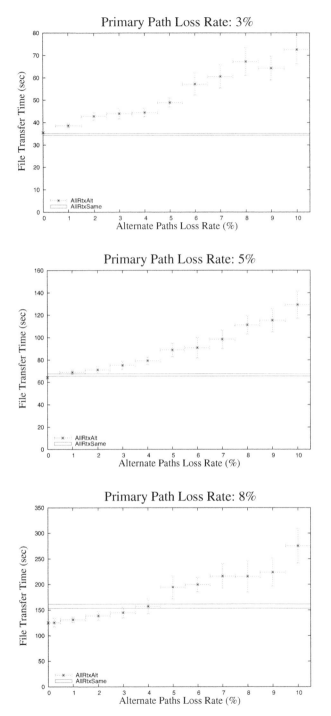

Figure 2.3 AllRtxAlt vs AllRtxSame at $\{3,5,8\}\%$ primary path loss.

with its an 8% loss rate. Clearly, this behavior is not what the SCTP authors expected when specifying the current retransmission policy.

Intuition tells us that when an alternate path's conditions are better than the primary's, then AllRtxAlt should improve performance, and when the conditions are worse on the alternate path, then AllRtxAlt should degrade performance. However, our results show that often the former expectation does not hold. Furthermore, independent results by other researchers confirm that AllRtxAlt degrades performance (Duke et al. 2003). The next section explains why.

Stale RTOs

After analyzing several experiment traces in detail, we attribute AllRtxAlt's poor performance to stale RTO values for the alternate path. Due to Karn's algorithm (Karn and Partridge 1987), successful retransmissions on the alternate path cannot be used to update the sender's RTT estimation of the alternate path. Timeouts on retransmissions, however, exponentially increase the RTO. The only traffic on the alternate path which updates the RTT estimate are the periodic heartbeat probes used to determine destination reachability, but these heartbeats are transmitted relatively infrequently (approximately every 30 seconds (Stewart et al. 2000). In many cases the RTO is exponentially increased more frequently than it can be reduced by an RTT estimate. The result is an overly conservative (i.e., too large) RTO on the alternate path for the majority of the association. Thus, anytime a retransmission on the alternate path is lost, a timeout occurs and the timeout is likely to be unnecessarily long. In addition, each timeout further contributes to the problem by doubling the RTO value.

Figure 2.4 illustrates the dynamics of the RTO values for the primary path (8% loss rate) and the alternate path (5% loss rate) during a 4MB file transfer using AllRtxAlt. This specific transfer sent a total of 2,889 original transmissions on the primary path, of which 229 had to be retransmitted on the alternate path, and of those retransmissions, 14 were lost and re-retransmitted back on the primary path. The RTO value of the primary path stays relatively low (average is 2.3 seconds) during most of the transfer, because successful new data transmission on the primary path updates the RTT estimation and reduces the RTO value (most likely back to the minimum of 1 second). On the other hand, the alternate path, even with a lower loss rate, maintains an average RTO value of 5.9 seconds – more than double the primary's. Figure 2.4's graph for the alternate path shows that the alternate path's RTO reduces only three times. In other words, only three heartbeats are successfully acked and used to measure and improve the alternate path's RTT estimate. The graph also shows seven timeouts exponentially increasing the RTO value of the alternate path.

2.2.2 Best of Both Worlds

We have demonstrated the tradeoffs between AllRtxAlt and AllRtxSame. AllRtxSame generally provides better performance, but AllRtxAlt may improve performance if the alternate path's loss rate is low enough to overcome the stale RTO problem. The difficulty in practice is that a sender may or may not have *a priori*

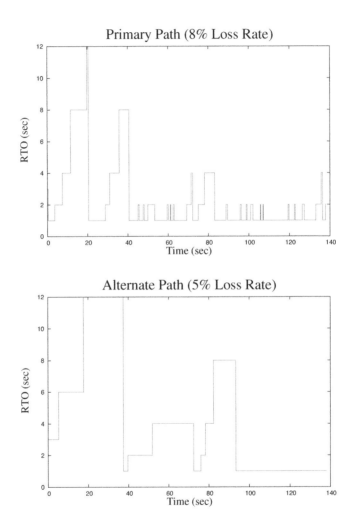

Figure 2.4 Example RTO dynamics with 8% primary path loss and 5% alternate path loss.

knowledge about the paths' conditions. Without such information, the best a sender can do is combine the good points of both policies. To do so, this author introduced a FrSameRtoAlt policy – a hybrid of AllRtxAlt and AllRtxSame. FrSameRtoAlt sends (a) fast retransmissions to the same destination as their original transmissions, and (b) timeout retransmissions to an alternate destination. Since timeouts tend to occur more often at higher loss rates, this policy increases the use of the alternate path as the primary path's loss rate increases. This section evaluates FrSameRtoAlt against AllRtxAlt and AllRtSame. We will determine whether FrSameRtoAlt does indeed effectively combine the good points of the other two policies.

Analysis Methodology

Figure 2.5 illustrates the network topology used, which is based on the topology previously presented in Figure 2.2. But instead of using cross-traffic to induce congestion-based loss, we introduce uniform loss on these paths (0-10% each way) at the core links. We realize that the cross-traffic approach used in Figure 2.2 more realistically simulates Internet traffic, but the simulation execution time for this technique became impractical. To evaluate if Figure 2.5's simplified model could still provide meaningful results, we compared representative simulations using the cross-traffic model from Figure 2.2 and the simpler uniform loss model from Figure 2.5. Although the absolute results differed for those examples compared, relative relationships remained consistent – leading to the same conclusions. We therefore proceeded with the simpler uniform loss model.

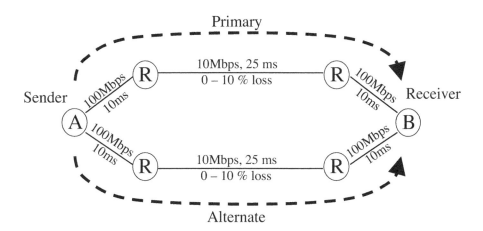

Figure 2.5 Simulation network topology with random loss.

The topology in Figure 2.5 maintains the same link bandwidths and delays used in Figure 2.2. The core links have 10Mbps bandwidth and 25ms one-way delay, and the edge links have 100Mbps bandwidth and 10ms one-way delay. Thus, the end-to-end RTT on either path is 90ms, which is a reasonable RTT within the continental United States.

We simulate a 4MB file transfer with three input parameters for each simulation: (1) the primary path's loss rate, (2) the alternate path's loss rate, and (3) one of the three retransmission policies. Each parameter set is simulated with 60 different seeds. We found 60 seeds to be sufficient for obtaining a 90% confidence interval.

Results

Figure 2.6 illustrates the results for $\{3, 5, 8\}\%$ primary path loss rates. For each graph in Figure 2.6, the alternate path's loss rate is varied on the x-axis, ranging from 0-10%. The graphs in Figure 2.6 compare the average transfer time of a 4MB file using each of the three policies: AllRtxAlt, AllRtxSame, FrSameRtoAlt.

We ensure statistical confidence by calculating the 90% confidence interval with an acceptable error of 10% of the mean. The 90% confidence intervals are not shown in the graphs for clarity. These intervals vary for different loss rates and retransmission policies, but on average the 90% confidence interval is about +/- 2-5 seconds around the mean. The largest 90% confidence interval is about +/- 13 seconds around the mean; as expected, larger confidence intervals occurred for higher loss rates and policies that use the alternate path more often.

Figure 2.6 clearly shows that (obviously) AllRtxSame's performance is uninfluenced by the alternate path's loss rate or by the stale RTO problem. Following the same trends observed in Section 2.2.1, the graphs in Figure 2.6 also show that AllRtxAlt may improve performance when the alternate path's loss rate is lower than the primary's, but the stale RTO problem dominates performance. First, AllRtxAlt does worse than AllRtxSame when both paths have the same bandwidth, delay, and loss rate. Second, the degree to which AllRtxAlt degrades performance is significantly higher than the degree to which it improves performance. For example, when the primary path loss rate is 5%, AllRtxAlt improves performance over AllRtxSame by 21% when the alternate path loss rate is 0%, but degrades performance by more than double (108%) when the alternate path loss rate is 10%.

FrSameRtoAlt, a hybrid policy, compromises between the advantages and disadvantages of AllRtxAlt and AllRtxSame. At low primary path loss rates (e.g., top graph in Figure 2.6), FrSameRtoAlt and AllRtxSame perform similarly. Most lost TPDUs at such loss rates are detected by the fast retransmit algorithm, and thus are retransmitted to the same destination. The relatively few timeouts that occur in these conditions are not enough to significantly influence the results.

As the primary path loss rate increases, AllRtxSame and FrSameRtoAlt begin to perform differently. An increase in the number of timeouts causes FrSameRtoAlt to send more traffic to the alternate destination. As a result, FrSameRtoAlt's performance depends more on the alternate path's loss rate. However, since FrSameRtoAlt does not send fast retransmissions to the alternate destination, the alternate path's loss rate influences FrSameRtoAlt's performance less than AllRtxAlt's. FrSameRtoAlt's improvements are not as great as AllRtxAlt's, but neither are the degradations. Fur-

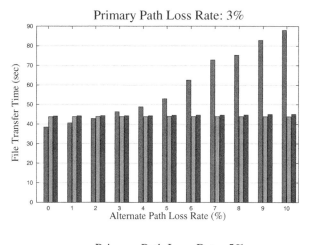

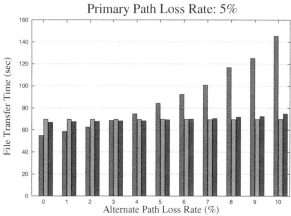

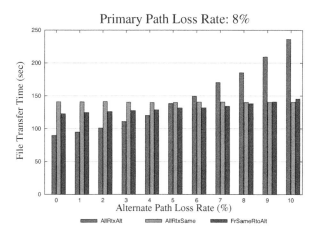

Figure 2.6   AllRtxAlt, AllRtxSame, and FrSameRtoAlt at {3,5,8}% primary path loss.

thermore, FrSameRtoAlt improves performance to a greater extent than it degrades performance. For example, when the primary path loss rate is 8%, FrSameRtoAlt improves performance over AllRtxSame by 13% when the alternate path loss rate is 0%, but degrades performance by only 3% when the alternate path loss rate is 10%. To contrast, AllRtxAlt offers a 36% improvement and 68% degradation under the same conditions.

When loss conditions of paths are unknown *a priori*, we need to consider overall performance. From the results in this section, we conclude that AllRtxAlt is the worst policy. AllRtxSame and FrSameRtoAlt perform about the same with FrSameRtoAlt offering a slight advantage when primary path loss rates are high.

2.2.3   Performance Enhancing Extensions

We now introduce three performance enhancing policy extensions. The motivation behind these extensions is to determine if the relative relationships between the retransmission policies remain unchanged even after trying to improve each one's performance.

Heartbeat after RTO (HAR)

When a timeout occurs, the Heartbeat After RTO (HAR) mechanism sends a heartbeat immediately to the destination on which a timeout occurred. This behavior, suggested by Randall Stewart, is in addition to the normal data retransmission behavior (specified by the retransmission policy) that remains unchanged. Since AllRtxSame only sends timeout retransmissions to the same destination, HAR is not applicable (see Figure 2.7). The extra heartbeats introduced by HAR try to eliminate the stale RTO problem of AllRtxAlt and FrSameRtoAlt. With HAR, a sender updates an alternate destination's RTT estimate more frequently, thus resulting in a better RTT estimate on which to base the RTO value, at the expense of extra traffic.

For example, suppose a TPDU is lost in transit to the primary destination, and later gets retransmitted to an alternate destination. Also suppose that the retransmission times out. The lost TPDU is retransmitted again to yet another alternate destination (if one exists; otherwise, the primary). More importantly, a heartbeat is also sent to the alternate destination which timed out. If the heartbeat is successfully acked, that destination acquires an additional RTT measurement to undo the exponentially backed off RTO.

Timestamps (TS)

The timestamp (TS) mechanism is similar to TCP's timestamp mechanism. By including timestamps in each TPDU, all retransmission ambiguity is resolved. That is, the sender can always determine which transmission (original or retransmission) an ack belongs to. Thus, Karn's algorithm can be eliminated, and successful retransmissions can be used to update a sender's RTT estimate, and in turn facilitate a more accurate RTO value. Using a timestamp mechanism in a multihomed transport protocol (introduced by this author) is especially useful in alleviating the stale RTO problem of AllRtxAlt and FrSameRtoAlt, but at the expense of a 12 byte overhead in each TPDU.

Note that this extension's motivation is to evaluate how much performance can be improved by eliminating the retransmission ambiguity problem. One alternative solution, incurring less TPDU overhead, may be to use flag(s) in the data and sack headers to signal whether the data/sack is for an original transmission or retransmission.

Multiple Fast Retransmit (MFR)

In TCP and the current SCTP specification, a TPDU is "fast retransmitted" after a sender receives a specified number of missing reports (known as duplicate acks, or dupacks for short, in TCP terminology). Currently, however, a TPDU can be fast retransmitted only once. The Multiple Fast Retransmit (MFR) algorithm, introduced by this author, introduces extra state at the sender to allow lost fast retransmissions, in some cases, to be fast retransmitted again instead of incurring a timeout. For example, suppose a sender has a window of data in flight to the receiver, and TPDU x is lost. Data successfully received at the receiver are sacked, and any sacks for TPDUs sent after x serve as missing reports for TPDU x. When the sender receives four such missing reports, the standard fast retransmit algorithm is triggered and TPDU x is retransmitted.[2] At this point, MFR state stores the highest TPDU currently outstanding, n. This way, if the retransmission of x is also lost, the sender can detect the loss with another four missing reports. Only sacks for TPDUs greater than n can serve as missing reports for another fast retransmission, because the sacks up to n were already in flight when x was fast retransmitted the first time.

MFR applies to AllRtxSame and FrSameRtoAlt. Since AllRtxAlt sends fast retransmissions to an alternate path, MFR could cause spurious fast retransmissions when path delays are different. For example, imagine a fast retransmission scenario where the primary path's RTT is shorter than the alternate path's. After a fast retransmission is sent on the alternate path, new data sent on the primary path may arrive at the receiver first. If so, the receiver uses sacks to convey this reordering to the sender. However, the sender's MFR algorithm will mistakenly interpret the reordering as loss of the fast retransmitted data, and incorrectly trigger another fast retransmission of the same data.

Although MFR can prevent some timeouts, it does not provide additional RTT samples for alternate destinations, and thus inevitable timeouts continue to suffer from the stale RTO problem. MFR may be combined with HAR or timestamps to address stale RTOs. Figure 2.7 illustrates all ten policy-extension combinations.

---

[2]SCTP (Stewart et al. 2000) requires four missing reports to trigger a fast retransmit, whereas TCP requires only three analogous dupacks (Allman et al. 1999).

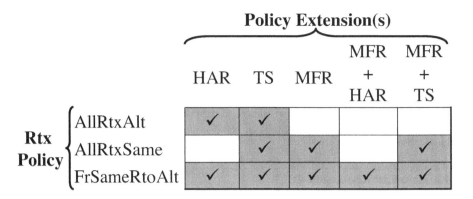

Figure 2.7 Possible policy-extension combinations.

Performance Evaluation

This section independently examines each policy, with its possible extensions. We determine which extension(s) provides the best improvement to each policy. For our evaluation, we use the methodology presented in Section 2.2.2.

AllRtxAlt's Extensions

Figure 2.8 presents the results for AllRtxAlt and its extensions with $\{3,5,8\}\%$ primary path loss rates. As the graphs show, both HAR and TS drastically improve AllRtxAlt's performance. HAR improves performance by as much as 38%, 43%, and 45% for primary path loss rates of 3%, 5%, and 8%, respectively. TS improves performance by slightly larger margins – as much as 45%, 51%, and 50%.

Both HAR and TS provide more RTT measurements of the alternate destination and reduce the occurrence of stale RTOs. Since HAR is a reactive mechanism that only obtains an extra measurement when timeouts occur, TS has an advantage over HAR. TS is proactive and offers more opportunities to measure the alternate path's RTT. Although TS adds a 12-byte overhead into each TPDU, the overhead does not significantly adversely impact performance. We conclude TS is the better extension for AllRtxAlt.

AllRtxSame's Extensions

Figure 2.9 presents the results for AllRtxSame and its extensions. Since AllRtxSame's performance is independent of the alternate path's conditions, we plot all the results in a single graph with the primary path's loss rate on the x-axis.

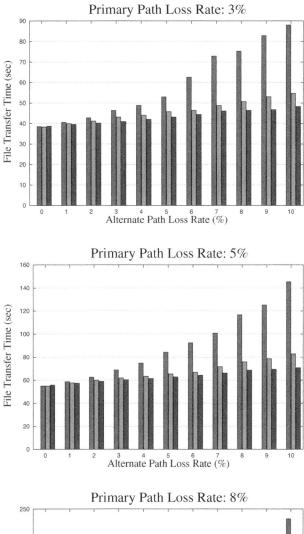

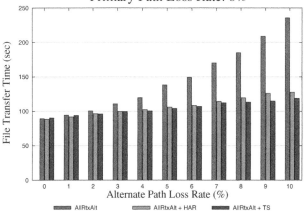

Figure 2.8 AllRtxAlt and its extensions at $\{3,5,8\}$% primary path loss.

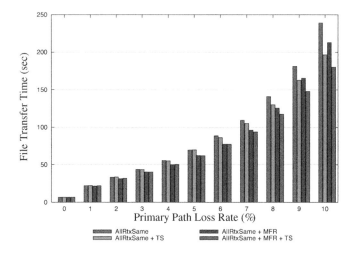

Figure 2.9 AllRtxSame and its extensions across all primary path loss rates.

The graph shows that MFR is able to avoid timeouts and increase AllRtxSame's performance. For example, MFR improves AllRtxSame's performance by 8%, 10%, and 11% under 3%, 5%, and 8% primary path loss rates. TS only improves performance when the primary path's loss rate is high. For example, including TS improves performance by 6-8% when the primary path's loss rate is 8%, but provides no benefit at 3% and 5% primary path loss. At high loss rates, timeouts may occur frequently enough that no RTT measurement is obtained between timeouts. Thus, TS improves performance by allowing a successful timeout retransmission to be used for measuring the RTT, which in turn decreases the exponentially backed-off RTO. Combining MFR and TS provides the best performance for AllRtxSame.

FrSameRtoAlt's Extensions

FrSameRtoAlt qualifies for five extension combinations, three of which include MFR. Figure 2.10 shows that individually, MFR provides greater improvement than either HAR or TS. Using HAR or TS alone, at best, provides 2%, 5%, and 9% improvement at 3%, 5%, and 8% primary path loss, respectively. MFR alone, on the other hand, improves performance by as much as 10%, 16%, and 14%. MFR's ability to avoid some timeouts has dramatic effects on FrSameRtoAlt's performance, because the stale RTO problem on the alternate path is also avoided.

Combining HAR or TS with MFR in general provides no added improvement; some marginal improvement occurs when the loss rate is high on the primary and alternate paths. For example, with 8% primary path loss and 10% alternate path loss, MFR+HAR and MFR+TS perform similarly and provide an additional 4-5%

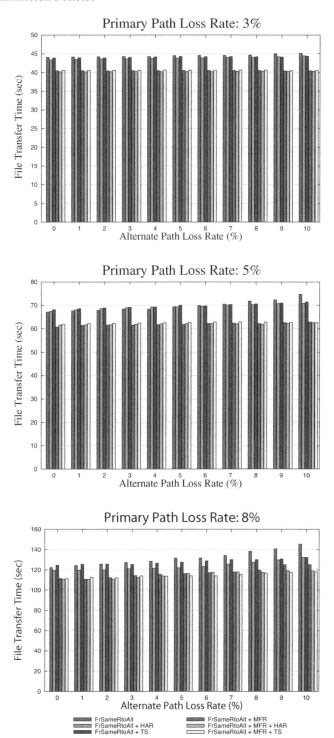

Figure 2.10 FrSameRtoAlt with its extensions at $\{3,5,8\}\%$ primary path loss.

improvement over MFR alone. Thus, FrSameRtoAlt performs best when combined with either MFR+HAR or MFR+TS. However, we recommend that MFR+TS be used, since TS (or any mechanism that eliminates the retransmission ambiguity) has other orthogonal uses that can improve performance, such as the Eifel algorithm (Ladha et al. 2004; Ludwig and Katz 2000).

2.2.4   Non-Failure Scenarios

This section revisits our performance comparison of the three policies in non-failure scenarios (done in Sections 2.2.1 and 2.2.2), but this time each policy is combined with our recommended extension(s): AllRtxAlt+TS, AllRtxSame+MFR+TS, and Fr-SameRtoAlt+MFR+TS. First, we evaluate their performance with symmetric path delays (i.e., both the primary and alternate paths have equal RTTs). Then, we assess the influence of the alternate path's delay in asymmetric trials. Finally, we consider three paths to determine if relative performance of the retransmission policies is influenced by the degree of multihoming. For readability throughout the remainder of this chapter, we refer to AllRtxAlt+TS, AllRtxSame+MFR+TS, and FrSameR-toAlt+MFR+TS as simply AllRtxAlt, AllRtxSame, and FrSameRtoAlt, respectively.

Analysis Methodology

We again use the methodology presented in Section 2.2.2 for our evaluation, but in this section we investigate alternate path RTTs. The primary path remains unchanged (see Figure 2.11). However, the alternate path's core link has three possible one-way delays: 25ms, 85ms, and 500ms (i.e., end-to-end RTTs of 90ms, 210ms, and 1040ms). These values sample reasonable RTTs experienced on the Internet. Although 1040ms may seem large, flows passing through cellular networks often experience RTTs as high as 1 or more seconds (Gurtov et al. 2002; Inamura et al. 2003; Jayaram and Rhee 2003).

Note that we do not simulate different link bandwidths. Lowering the alternate path's bandwidth simply increases the RTT, which we already independently control. Thus, the bandwidths remain constant in all our simulations.

Symmetric Path Delays

Figure 2.12 illustrates the results for $\{3,5,8\}\%$ primary path loss rates, a 90ms primary path RTT, and a 90ms alternate path RTT. Our first observation is that the extensions reduced the performance gap between the three retransmission policies (compare Figure 2.12 with Figure 2.6). For 3% primary path loss, the three policies perform relatively the same (less than 5% difference) for 0-4% alternate path loss. Higher alternate path loss rates cause AllRtxAlt to degrade performance by as much as 20%, while the results for AllRtxSame and FrSameRtoAlt remain unchanged.

When the primary path loss rate is 5%, AllRtxSame and FrSameRtoAlt again perform similarly. AllRtxAlt, on the other hand, improves performance by as much as 10% and degrades performance by as much as 14%, depending on the alternate

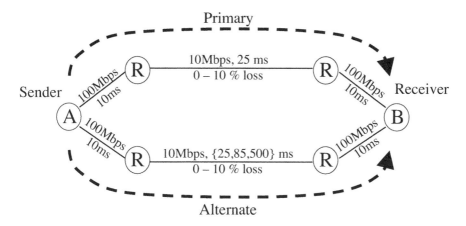

Figure 2.11 Simulation network topology with random loss, 90ms primary path RTT, and {90, 210, 1040}ms alternate path RTT.

path's loss rate (generally an unknown metric). Comparing this relatively low degradation to the degradation of 108% presented in Section 2.2.2 for the same network conditions, the stale RTO problem seems to have been completely eliminated.

The results for 8% primary path loss further confirm this observation. AllRtxAlt and FrSameRtoAlt outperform AllRtxSame across nearly all alternate path loss rates, and do about the same (within 2% of each other) when that alternate path loss rate is higher than 8%.

Overall, the results in Figure 2.12 do not present a decisive argument for a particular best policy. FrSameRtoAlt outperforms AllRtxSame, but deciding between AllRtxAlt and FrSameRtoAlt is not straightforward. FrSameRtoAlt provides only conservative gains, but does not degrade performance at all. AllRtxAlt may provide more significant gains, but risks the potential of degradation of the same magnitude.

Asymmetric Path Delays

We find that increasing the alternate path's RTT to slightly more than double (210ms) does not significantly affect performance. Although the results are not shown, the graphs are similar to those in Figure 2.12. Hence, we push the limits further and present in Figure 2.13 the performance when the alternate path's RTT is more than ten times longer than the primary.

The most obvious result is that AllRtxAlt's heavy use of the alternate path significantly degrades performance when the alternate path delay is large (no surprise). AllRtxSame's performance remains unchanged, as expected. FrSameRtoAlt's results, however, prove interesting. At 3% primary path loss, few timeouts occur. Hence, the alternate path is rarely used and FrSameRtoAlt's results remain unchanged. With a 5% and 8% primary path loss rate, FrSameRtoAlt degrades performance compared

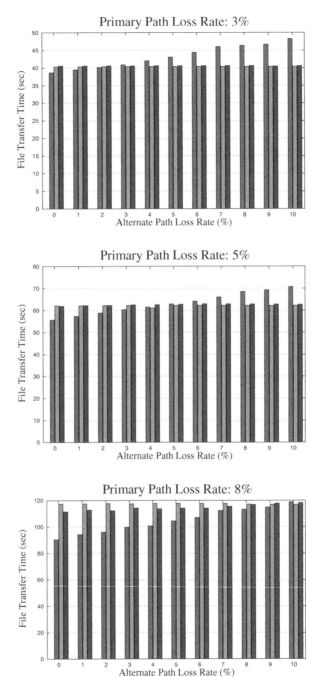

Figure 2.12 AllRtxAlt+TS, AllRtxSame+MFR+TS, and FrSameRtoAlt+MFR+TS at {3,5,8}% primary path loss, 90ms primary path RTT, and 90ms alternate path RTT.

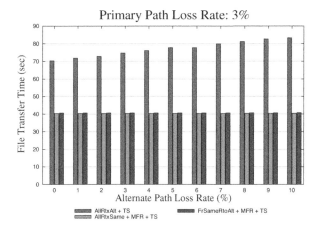

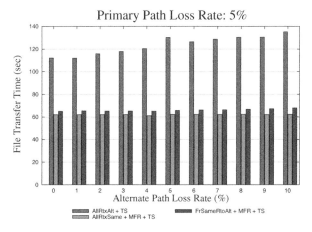

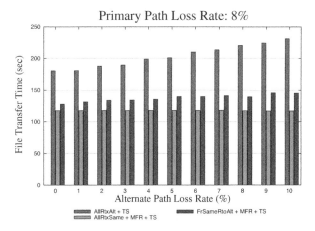

Figure 2.13 AllRtxAlt+TS, AllRtxSame+MFR, and FrSameRtoAlt+MFR+TS at $\{3,5,8\}\%$ primary path loss, 90ms primary path RTT, and 1040ms alternate path RTT.

to AllRtxSame, but given the large difference in path delays, this degradation is minor. The alternate path's delay is more than ten times that of the primary, but in the worst case, FrSameRtoAlt degrades performance by only 9% and 24% for primary path loss rates of 5% and 8%, respectively.

Therefore, if the alternate path delay is *known* to be an order of magnitude longer than the primary's, we suggest using AllRtxSame. Otherwise, the conclusions from Section 2.2.4 remain.

Three Paths

To determine if our conclusions hold when the number of paths between the endpoints increases, we add an additional alternate path to the topology in Figure 2.11. We configure both alternate paths to have the same properties (bandwidths, delays, and loss rates) as each other. Otherwise, the number of simulation parameters would quickly become unmanageable. The results (not shown) are similar to those for two paths. That is, the relationships between the policies remain the same. We expect that the trends will remain the same for configurations with more than three paths between endpoints.

2.2.5   Failure Scenarios

We again evaluate the performance of the three policies with their best performing extension(s), this time focusing on failure scenarios, an important criteria in the overall evaluation. After all, a key motivation for supporting multihoming at the transport layer is improved failure resilience. Hence, a multihomed transport layer should use a retransmission policy that performs well when the primary destination becomes unreachable.

Failover Algorithm

Each endpoint uses both implicit and explicit probes to dynamically maintain knowledge about the reachability of its peer's IP addresses. Transmitted data serve as implicit probes to a destination (generally, the primary destination), while explicit probes, called *heartbeats*, periodically test reachability and measure the RTT of idle destinations. Each timeout (for data or heartbeats) on a particular destination increments an error count for that destination. The error count per destination is cleared whenever data or a heartbeat sent to that destination is acked. A destination is marked as failed when its error count *exceeds* the failover threshold (called Path.Max.Retrans in SCTP).

If the primary destination fails, the sender fails over to an alternate destination address, and continues probing the primary destination with heartbeats. Failover is temporary in that a sender resumes sending new data to the primary destination if and when a future probe to the primary destination is successfully acked. If more than one alternate destination address exists, RFC2960 leaves the alternate destination selection method unspecified. We assume a round-robin selection method.

RFC2960 recommends default settings of: minimum RTO = 1s, maximum RTO = 60s, and Path.Max.Retrans (PMR) = 5. Using these defaults, the first timeout towards failure detection takes 1s *in the best case*. Then, the exponential back-off procedure doubles the RTO on each subsequent timeout towards failure detection. With PMR = 5, six consecutive timeouts are needed to detect failure, taking at least $1+2+4+8+16+32 = 63$s. In the worst case, the first timeout takes the maximum of 60s, and the failure detection time requires $6*60 = 360$s.

Analysis Methodology

We use the same methodology described in Section 2.2.2, but in this section we introduce failure scenarios. The topology in Figure 2.14 shows that the both paths have the same characteristics, except that some time during the file transfer, the primary path's core link experiences a bi-directional failure. We simulate a link breakage between the routers on the core path at two different times. The first set of failure experiments experience the link breakage at time = 4s into the transfer. With 0% loss on the primary path, roughly half of the 4MB file transfer is complete by this time. The second set's link breakage occurs at time = 6.8s into the transfer, diabolically chosen to occur during the last RTT of the 4MB file transfer when the primary path's loss rate is 0%. In both failure scenarios, the link remains down until the end of the simulation.

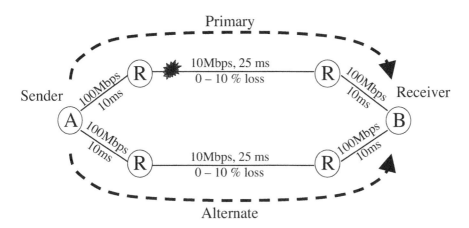

Figure 2.14 Simulation network topology with random loss, equal delays, and primary path failure.

Results

To gauge the performance during failure scenarios, we not only measure the file transfer time, we also consider the timeliness of data. File transfer in a failure scenario can be divided into three periods: (1) before failure, (2) during failure detection, (3) after failover. The first period has been covered in Section 2.2.4.

The second period, failure detection, is important for both file transfer time and data timeliness. Fast failover time improves file transfer time, because the sender is able to resume "normal" transmission on an alternate path more quickly. As expected, we find that the failure detection time is similar for the three retransmission policies.

The retransmission policy affects timeliness of data in that it determines whether a transfer is stalled during the failure detection process. AllRtxSame delivers no data to the peer until the entire failure detection process completes and failover occurs. For example, with 0% primary path loss, the sender has 30 lost data TPDUs outstanding when failure occurs in our first failure scenario (link breakage at time = 4s). AllRtx-Alt and FrSameRtoAlt successfully retransmit these 30 TPDUs after the first timeout in the failure detection process, thus delaying them by only 1s (or whatever the primary path's RTO is at that point). Furthermore, during each subsequent timeout that contributes to failure detection, the sender successfully retransmits one TPDU to the alternate destination. On the other hand, with AllRtxSame the sender successfully retransmits the initial 30 lost TPDUs only after the failure detection completes, delaying them by at least 63s! While perhaps not an issue for a file transfer, as being simulated in our experiments, this delay may be unacceptable to applications requiring timely or consistent rate-based data delivery.

During the third period, the sender has only one available path for transmission in our simulations. (The results in Section 2.2.4 apply to scenarios where more than one path are available during the third period.) Figure 2.15 presents the final transfer times for failure at time = 4s. As the graphs show, the primary path's loss rate has minimal influence on the file transfer time. Comparing these results with those in Figure 2.12 suggests that the third period has heaviest influence on file transfer time. Since failure occurs relatively early in the file transfer, the remaining portion of the transfer is large enough that its sole use of the alternate path is the most influential factor on file transfer time. Even the policies themselves do not provide much difference (at most 9%) in performance. Since there is only one available path in the third period, all three retransmission policies perform similarly, differing only by the extensions used.

As a worst case example, the file transfer times for 0% primary path loss and failure at time = 6.8s are shown in Figure 2.16. Since this failure scenario has a link breakage in the last RTT of the data transfer, the second period (i.e., failure detection) is the most influential factor on file transfer time. Figure 2.16 shows that AllRtxSame's transfer stalls during failure detection has drastic effects on the results. The file transfer with AllRtxSame requires complete failure detection and takes about 70s to complete, whereas it only takes about 8-18s (depending on the alternate path's loss rate) with AllRtxAlt and FrSameRtoAlt. The reason is that AllRtxSame is unable

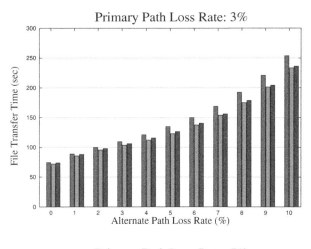

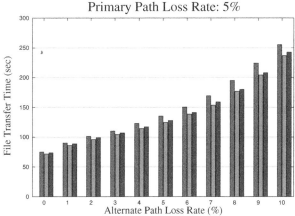

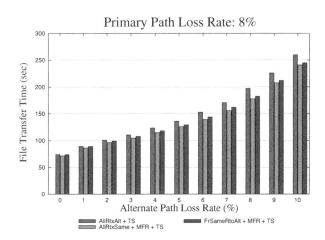

Figure 2.15 AllRtxAlt+TS, AllRtxSame+MFR+TS, and FrSameRtoAlt+MFR+TS with primary path failure at time = 4s.

to complete the transfer until after failover occurs, but AllRtxAlt and FrSameRtoAlt are able to finish the transfer during failure detection. Note that this example indeed represents a worst case situation for AllRtxSame, and was diabolically conceived.

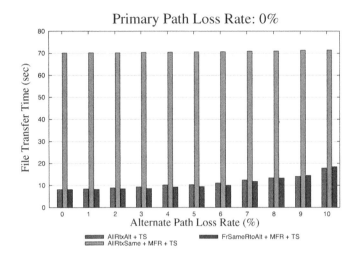

Figure 2.16 AllRtxAlt+TS, AllRtxSame+MFR+TS, and FrSameRtoAlt+MFR+TS with primary path failure at time = 6.8s.

In summary, all three policies provide similar throughput performance for large transfers during failure scenarios. However, AllRtxSame's transfer stalls during failure detection degrades performance if the failure coincidentally occurs near the end of the transfer and/or data timeliness is important. Hence, AllRtxAlt and FrSameRtoAlt are recommended for failure scenarios.

2.2.6   Summary

We have evaluated three retransmission policies for multihomed transport protocols, using SCTP to demonstrate the concepts. Without *a priori* knowledge about the available paths, a sender cannot have a static policy that decides where to retransmit lost data and expect to guarantee the best performance. Through simulation, we have measured and demonstrated the tradeoffs of three policies in non-failure and failure conditions. Our results show that the retransmission policy which best balances the tradeoffs is this author's hybrid policy: (1) send fast retransmissions to the same peer IP address as the original transmission, and (2) send timeout retransmissions to an alternate peer IP address.[3] We have shown that this hybrid policy performs best when combined with two enhancements: our Multiple Fast Retransmit algorithm,

---

[3]This policy has been proposed to the IETF as a change to SCTP (Stewart et al. 2005a).

and either timestamps or our Heartbeat After RTO mechanism. The Multiple Fast Retransmit algorithm reduces the number of timeouts. Timestamps and the Heartbeat After RTO mechanism both improve performance when timeouts are common by providing extra RTT measurements and maintaining low RTO values – an important feature for alternate paths that are mostly idle.

2.3 Failover Thresholds

SCTP has a tunable failover threshold that RFC2960 recommends should be set to a conservative value of six consecutive timeouts, which translates to a failure detection time of at least 63 seconds – unacceptable for many applications. In this section, we measure the tradeoff between more aggressive failover (i.e., lower thresholds) and spurious failovers. Lower thresholds provide faster failure detection when failure has occurred, but cause spurious failovers in non-failure lossy conditions. However, we surprisingly find that spurious failovers do not degrade performance, and often actually improve goodput regardless of the paths' characteristics (bandwidth, delay, and loss rate).

Currently, traffic migrates back to the primary path when the primary path recovers (i.e., failovers are temporary). This temporary nature throttles the sending rate, because upon returning to using the primary path, the sender must enter slow start with a cwnd of one MTU. To avoid this slowdown, this author introduces the concept of permanent failovers in multihoming protocols, where a sender makes the failover permanent if the primary path does not respond within some threshold amount of time. We find that permanent failovers can improve performance when a sender has estimates of each path's RTT and loss rate to make an informed decision. In the case a sender does not have such information, we recommend that permanent failovers not be used.

Section 2.3.1 describes SCTP's current failover mechanism. Section 2.3.2 presents the tradeoffs between more aggressive failover and spurious failovers. Section 2.3.3 introduces and evaluates a modified failover mechanism that allows failovers to become permanent. We summarize our failover results in Section 2.3.4.

2.3.1 SCTP's Failover Mechanism

Section 2.2.5 presented the details of the failover algorithm, but here we present its formal specification, as shown in Figure 2.17 for n destinations. The association begins in Phase I, where destination D_i is the primary destination, D_i is in the active state, and all new data are sent to D_i. When D_i fails, "failover" occurs and the association moves into Phase II.

In Phase II, D_i remains the primary destination, but in a failed state. All new data are redirected to an alternate destination, D_j. If more than one alternate destination address exists, RFC2960 leaves the alternate destination selection method unspecified. In this work, we assume a round-robin selection method. If D_j's error count

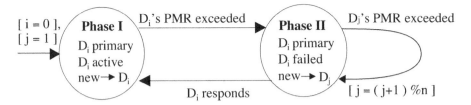

Figure 2.17 FSM for current failover mechanism.

should exceed PMR, a failover occurs to yet another alternate destination and the association stays in Phase II.

While in Phase II, the sender explicitly probes the primary destination, D_i, with periodic heartbeats. If D_i ever responds (i.e., recovers), failover is cancelled and the association returns to Phase I.

2.3.2   Reducing PMR

Reducing PMR decreases failure detection time, but increases the possibility of *spurious failover*, where a sender concludes a failure has occurred (when in fact timeouts were due to congestion). In this section, we measure the tradeoff between lower PMR settings and spurious failovers. The goal is to determine how much failure detection time can be improved without having detrimental effects on goodput.

Methodology

Figure 2.18 illustrates the network topology used to evaluate different PMR settings. The multihomed sender, *A*, has two paths (labeled *Primary* and *Alternate*) to the multihomed receiver, *B*. The primary path's core link has a 10Mbps bandwidth and a 25ms one-way delay. The alternate path's core link has a 10Mbps bandwidth and one-way delays of 25ms, 85ms, and 500ms. Each router, *R*, is attached to a dual-homed node (*A* or *B*) via an edge link with 100Mbps bandwidth and 10ms one-way delay.

The end-to-end RTTs are 90ms, 210ms, and 1040ms, which sample reasonable delays on the Internet today. An RTT of 1040ms may seem large, but as state in Section 2.2.4, flows passing through cellular networks often experience RTTs as high as 1 or more seconds (Gurtov et al. 2002; Inamura et al. 2003; Jayaram and Rhee 2003). In any case, the delays are selected to demonstrate relative performance, and we believe our results and conclusions are independent of the actual bandwidth and delay configurations.

Note that we do not simulate different link bandwidths. Reducing the alternate path's bandwidth simply increases the RTT, which we already independently control.

We introduce uniform loss on these paths (0-10% each way) at the core links. Again, we realize that using cross-traffic to cause congestion would more realisti-

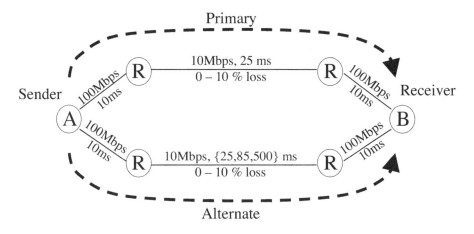

Figure 2.18 Simulation network topology.

cally simulate loss, but we found the simulation time for such a technique became impractical. On the other hand, uniform loss is a simple, yet sufficient model to provide insight about the effectiveness of different PMR settings accurately detecting failure. To evaluate if Figure 2.18's loss model was reasonable, we compared representative simulations using a cross-traffic model, shown in Figure 2.2, to produce self-similar, bursty traffic. Although the absolute results differed for those examples compared, relative relationships remained consistent – leading to the same conclusions. We therefore proceeded with the simpler uniform loss model.

In our simulations, the sender uses a different retransmission policy than specified in RFC2960. Instead, the sender transmits using this author's hybrid policy, FrSameRtoAlt, which was shown in Section 2.2 to perform better for *roughly* asymmetric path delays (i.e., no path delay is more than double of another). In our simulations, the Multiple Fast Retransmit algorithm (see Section 2.2.3) is also used to reduce the number of timeouts.

To observe long term averages, we simulate 80MB file transfers with PMR = $\{0, 1, 2, 3, 4, 5\}$. In this study, no link or interface failures are introduced; hence, all failovers that do occur are spurious. Each simulation has four parameters:

1. primary path's loss rate

2. alternate path's loss rate

3. alternate path's core link delay

4. PMR setting

Spurious Failovers

Figure 2.19 plots, for each PMR setting, the fraction of 80MB file transfers that experience at least one spurious failover at primary path loss rates 0-10%. The graph aggregates all alternate path loss rates for each particular primary path loss rate.

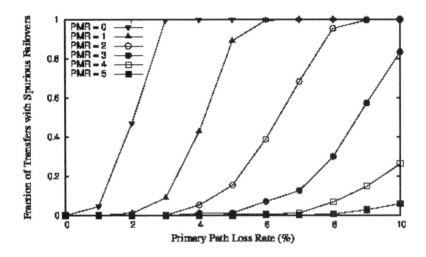

Figure 2.19 Fraction of transfers with spurious failovers.

Since PMR = 0 triggers a failover on a single timeout, this setting provides little robustness against spurious failovers at loss rates greater than 1%. At the other extreme, PMR = 5 experiences nearly no spurious failovers at loss rates less than 8%. As PMR increases from 0-5, the corresponding curves shift to the right by a loss rate of about 2%. This trend implies a simple linear relationship between the PMR setting and the robustness against spurious failovers. However, the slopes of the curves slowly flatten as the PMR increases, which argues that the robustness increases by more than a constant for each PMR setting.

The frequency of spurious failovers is also important when considering the robustness of various PMR settings. Figure 2.20 plots the cumulative distribution function (CDF) of the number of spurious failovers for primary path loss rates 2-10%. The CDFs for 1% primary path loss rate are omitted, because PMR = {1,2,3,4,5} experience no spurious failovers, and PMR = 0 experiences spurious failovers in only 5% of the transfers. Again, each graph in Figure 2.20 aggregates all alternate path loss rates for each primary path loss rate.

At a 2% primary path loss rate, 53% of transfers with PMR = 0 experience no spurious failovers, and 84% of transfers spuriously failover at most once. When the loss rate increases to 3%, less than 1% of transfers with PMR = 0 experience no spurious failovers. Then with 4% loss, only 1% of transfers experience less than ten spurious failovers.

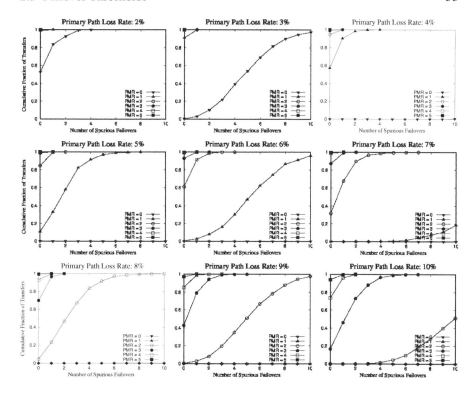

Figure 2.20 CDF of the number of spurious failovers for primary path loss rates 2-10%.

As expected, PMR = 1 is more robust against spurious failovers than PMR = 0. At 3% loss, 91% of the transfers do not spuriously failover. Furthermore, at 4% loss, 57% of the transfers are free of spurious failovers, and no transfers experience more than four failovers. When the loss rate is 8%, less than 1% of transfers observe less than ten spurious failovers.

This trend continues for PMR = {2, 3, 4, 5}. More than 25% of the transfers observe spurious failovers at {6, 8, 10}% loss for PMR = {2, 3, 4}. With PMR = 5, only 3% and 6% of transfers have spurious failovers at 9% and 10% loss, respectively.

To conclude, determining which failover threshold is "robust enough" largely depends on the networking environment. For example, Zhang et al. (2001) use end-to-end Internet measurements to report that 84% of their traces experienced less than a 1% loss rate (i.e., essentially "lossless"), and 15% of their traces had loss rates of 1-10% (with an average of 4%). Thus, to be completely robust against spurious failovers on 99% of Internet paths, PMR should be set to 6 (even PMR = 5 spuriously fails over 6% of the time at 10% loss), but that translates to a failover time of 123 seconds! *However, we conclude that PMR = 3 is robust enough for the Internet. This setting translates to a 15 second failover time, and is robust for all "lossless" paths and the average "lossy" path.*

Symmetric Path Delays

While the frequency of spurious failovers is important in providing intuition about overall behavior, of greater importance is how these spurious failovers affect performance. We collected results for 0-10% loss on the primary and alternate paths, but we do not include all results. Figure 2.21 plots the average 80MB file transfer time for $\{3, 5, 8, 10\}$% primary path loss, a 90ms primary path RTT, and a 90ms alternate path RTT. Each graph has a fixed primary path loss rate, and varies the alternate path loss rate on the *x*-axis from 0-10%.

Counter to our intuition, we observe that the PMR setting has little effect on the goodput for primary path loss rates less than 8%. Above 8%, the results show that lower (!) PMR settings begin to improve performance, with PMR $= 0$ providing the most improvement. That is, surprisingly, being more aggressive with failover often provides improved performance, even when the alternate path loss rate is higher than that of the primary path. For example, reducing the PMR from 5 to 0 improves the performance by 4% when the primary and alternate path loss rates are 8% and 10%, respectively. These counter-intuitive results will be explained later in Section 2.3.2.

Asymmetric Path Delays

We are also surprised to find that being aggressive with failover does not change with asymmetric path delays. We expected lower PMR settings to perform relatively worse when the alternate path has a larger RTT than the primary path. However, we find that the results remain nearly constant regardless of the alternate path delay. Figure 2.22 plots the results for $\{3, 5, 8, 10\}$% primary path loss, a 90ms primary path RTT, and a 1040ms alternate path RTT. Comparing these results with those in Figure 2.21 shows that the alternate path's longer RTT does not affect the performance. Even when the alternate path's RTT is more than ten times longer than the primary path's, PMR $= 0$ outperforms other PMR settings. Again, these results were unexpected, and will be explained in Section 2.3.2.

Three Paths

To determine if our conclusions hold when the number of paths between the endpoints increases, we add an additional alternate path to the topology in Figure 2.18. We configure both alternate paths to have the same properties (bandwidths, delays, and loss rates) as each other. Again, we want to avoid an unmanageable number of simulation parameters. The results (not shown) are consistent with those for two paths. That is, the relationships between the different PMR settings remain the same. We expect that the trends will remain the same for configurations with more than three paths between endpoints.

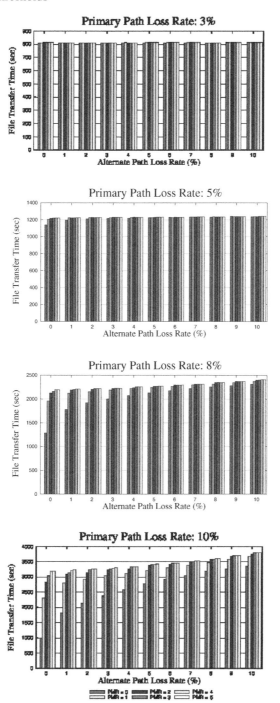

Figure 2.21 PMR evaluation: 90ms primary path RTT and 90ms alternate path RTT.

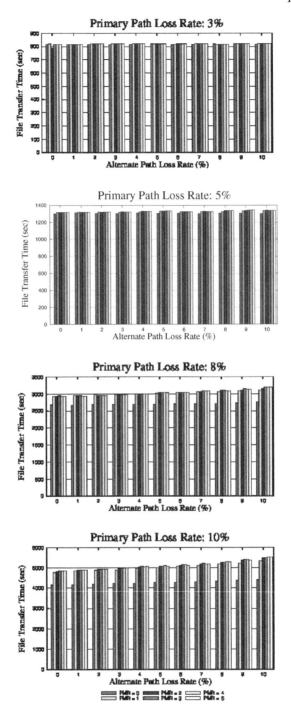

Figure 2.22 PMR evaluation: 90ms primary path RTT and 1040ms alternate path RTT.

Dormant State Behavior

As the finite state machine in Figure 2.17 shows, if a sender fails over to an alternate destination that in turn fails, the sender will failover to yet another alternate destination. If needed, the sender continues to failover to other alternate destinations until all alternate destinations are exhausted. When all destinations have failed, the association enters the *dormant state* (Stewart and Xie 2001) (not represented in Figure 2.17).

RFC2960 does not specify dormant state behavior. Implementations are provided the freedom of choosing what action a sender takes when all destinations fail. The association leaves the dormant state when one of the destinations (primary or alternate) responds. Otherwise, the association is aborted when the association exceeds the Association.Max.Retrans threshold, which is an SCTP parameter to limit the number of consecutive timeouts across all destinations.

In the current SCTP specification, dormant state behavior is considered unimportant, because high PMR settings make the dormant state unlikely. However, if PMR is lowered to 0, as our results thus far argue should be done, entering the dormant state becomes more likely. Thus, this author has identified three different dormant state behaviors to evaluate how they might impact performance: (1) Dormant LastDest, (2) Dormant Primary, and (3) Dormant Hop.

The Dormant LastDest behavior dictates that when the dormant state is entered, the sender continues sending new data to whichever destination was last used in Phase II. The other destinations still are periodically probed in the background with heartbeats. If the primary destination replies, the dormant state is exited, and the association returns to Phase I (permanent failover will be introduced in Section 2.3.3). If an alternate destination replies, the association returns to Phase II with the destination that replied as D_j.

The Dormant Primary behavior differs only slightly from the Dormant LastDest behavior. Instead of continually sending new data to whichever destination was last used in Phase II, the sender continually sends new data to the primary destination.

The Dormant Hop behavior, shown in Figure 2.23, attempts to be more aggressive in finding an active destination. While in the dormant state, the sender transmits new data to a different destination after each timeout. The sender cycles through all the destinations in a round-robin fashion until either a destination responds, or the association aborts.

The results in Sections 2.3.2 through 2.3.2 use the Dormant Hop behavior, but we also evaluate the performance of the other two dormant state behaviors. We find that dormant state behavior does not affect goodput, and the trend reported in those sections remains consistent for all dormant state behaviors (results not shown).

Analysis

Our results document that aggressive failover settings (in particular, PMR $= 0$) improve performance regardless of the path loss rates, path delays, and/or dormant state behavior – a result counter to our intuition. We spent considerable time investigating

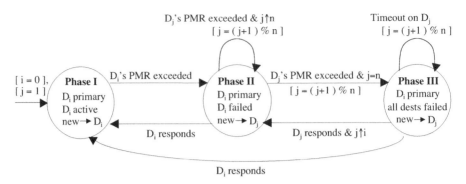

Figure 2.23 FSM with Dormant Hop behavior.

this surprising conclusion, which we now explain.

The underlying advantage of aggressive failover is that an association spends less time stalled during failure detection. With PMR = 0 for example, a single timeout moves new data transmission to the alternate path while the primary destination is probed with heartbeats. The primary destination may respond on the first probe, or it may not respond for a long time. In either case, data transmission continues on the alternate path, and migrates back to the primary path if and when the primary destination responds. Less aggressive failover settings (e.g., PMR = 5) cause a sender to wait longer before sending new data to the primary destination; in the meantime, essentially no useful communication takes place. Therefore, even if the alternate path has a higher loss rate and/or longer RTT, the sender always has the potential to gain (without risking doing worse) by failing over sooner.

The remainder of this section presents four detailed timeout scenarios (shown in Figure 2.24) for PMR = {0, 1} to demonstrate the merits of more aggressive failover. They all begin with TSN 1 (i.e., TPDU 1) being lost in transit to the primary destination and subsequently timing out. For PMR = 0, the sender immediately fails over, retransmits TSN 1 to the alternate destination, and sends a heartbeat to the primary destination. For PMR = 1, the sender retransmits TSN 1 to the alternate destination and sends TSN 2 to the primary destination. We compare the behavior of these two PMR settings by following the details of four (of many) possible scenarios beyond this point.

Scenario 1

The first TPDU sent to the primary destination and the first TPDU sent to the alternate destination following TSN 1's timeout are both delivered successfully.

- **PMR = 0** The failover is cancelled when the heartbeat is acked. Although the figure shows both TSN 1 and the heartbeat are acked at the same time, it is

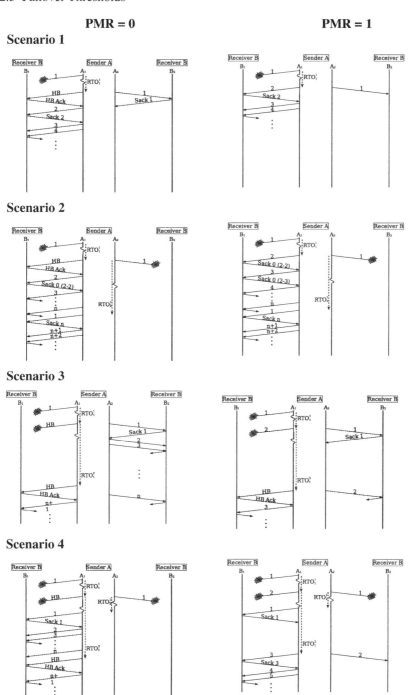

Figure 2.24 Timeout scenarios.

a race condition. If the heartbeat gets acked first (as shown in Figure 2.24's Scenario 1), then TSN 2 is sent on the primary and normal data transfer continues from this point. If TSN 1 gets acked first (not shown), then TSNs 2-3 are sent to the alternate destination, TSN 4 is sent to the primary destination when the heartbeat is acked, and normal data transfer continues to the primary destination.

- **PMR = 1** As both TSN 1 and 2 are sent at about the same time, again a race condition occurs. If TSN 1 arrives at the receiver first, the receiver's delayed ack algorithm causes a single cumulative ack (denoted SACK 2) to be generated for both TSN 1 and 2 (as shown in Figure 2.24). When this ack arrives, TSNs 3-4 are sent to the primary destination and normal data transfer continues to the primary destination. If TSN 2 arrives at the receiver first, the receiver generates two acks (not shown). The first selectively acks TSN 2 with a missing report for TSN 1, and the second cumulatively acks TSN 2. Upon receiving the first, the sender sends TSN 3 to the primary destination and normal data transfer continues to the primary destination.

This scenario presents a case where both PMR settings perform *roughly* similar in our experiments. Let RTT_1 and RTT_2 be the primary path's RTT and the alternate path's RTT, respectively. If $RTT_1 \leq RTT_2$ (as is the case in our experiments), then PMR $= 1$ has a marginal advantage in that it sends one more TPDU than PMR $= 0$.

On the other hand, if $RTT_1 > RTT_2$ (not shown in Figure 2.24 and not the case in our experiments), then PMR $= 0$ gets ahead of PMR $= 1$ in the overall transfer. The amount by which PMR $= 0$ gets ahead depends on the ratio of the two paths' RTTs. However, since $RTT_1 \leq RTT_2$ in our experiments, we omit detailed analysis of PMR $= 0$'s performance gain when $RTT_1 > RTT_2$.

Scenario 2

The first TPDU sent to the primary destination following TSN 1's timeout is successfully delivered, and the first TPDU sent to the alternate destination is lost.

- **PMR = 0** The failover is cancelled when the heartbeat is acked. TSN 2 is sent to the primary destination. When TSN 2 is selectively acked, TSN 3 is then sent to the primary destination. The sender continues sending one TPDU at a time to the primary destination until TSN 1's retransmission times out. TSN 1 is then re-retransmitted to the primary destination and normal data transfer continues to the primary destination.

- **PMR = 1** When TSN 2 is selectively acked, TSN 3 is sent to the primary destination, and when it is selectively acked, TSN 4 is sent to the primary destination. The sender continues sending one TPDU at a time to the primary destination until TSN 1's retransmission times out. TSN 1 is then re-retransmitted to the primary destination and normal data transfer continues to the primary destination.

Again, both PMR settings perform *roughly* similar. PMR = 1 has only a marginal advantage in that it sends one more TPDU than PMR = 0. This scenario shows that loss on the alternate path alone has little effect on the performance gap between PMR settings.

Scenario 3

The first TPDU sent to the primary destination following TSN 1's timeout is lost, and the first TPDU sent to the alternate destination is delivered successfully.

- **PMR = 0** When TSN 1 is acked, TSNs 2-3 are sent to the alternate destination, and normal data transfer continues temporarily to the alternate destination. Eventually, the heartbeat times out, and another heartbeat is then sent to the primary destination. Since this timeout is the second consecutive timeout on the primary destination, it will take at least 2 seconds to expire (assuming RTO.Min is 1 second). Once the second heartbeat is successfully acked, the sender cancels the failover, and resumes normal data transmission to the primary destination.

- **PMR = 1** When TSN 1 is acked, the sender is temporarily stalled and does not send any new data. When TSN 2 times out (again, at least 2 seconds later), the sender fails over to the alternate destination, retransmits TSN 2 to the alternate destination, and sends a heartbeat to the primary destination. From this point, normal data transfer continues to the alternate destination until the heartbeat is acked and the failover is cancelled. Then the sender resumes normal data transfer to the primary destination.

In Scenario 3, PMR = 0 may potentially perform significantly better than PMR = 1. With PMR = 0, the sender transmits new data on the alternate path until the sender receives a heartbeat ack from the primary destination. We estimate the number of TPDUs, d, sent to the alternate destination during this period as follows. From the time TSN 1 is retransmitted, the time it takes to receive a heartbeat ack from the primary destination is $(RTO_1^2 + RTT_1)$, where RTO_1^2 is the primary path's RTO for the second consecutive timeout, and RTT_1 is the primary path's RTT. The number of alternate path round trips, r, that will take place during this period is

$$r = min\left[1, \frac{RTO_1^2 + RTT_1}{RTT_2}\right] \tag{2.1}$$

where RTT_2 is the alternate path's RTT. Note that since at least one TPDU (TSN 1) is successfully sent on the alternate path, r must be at least 1.

To estimate the number of TPDUs, d, sent to the alternate destination during r alternate path round trips, we first assume that no loss occurs on the alternate path during this period. Hence, the transfer on the alternate path exits slow start when cwnd exceeds ssthresh. Using the slow start cwnd growth model from Cardwell et al. (2000), the last alternate path cwnd before exiting slow start is

$$cwnd = ssthresh = init\_cwnd \cdot \left(1 + \frac{1}{b}\right)^{r_{ss}-1} \tag{2.2}$$

where $init\_cwnd$ is the initial cwnd, b is the number of TPDUs per ack the receiver's delayed ack algorithm uses, and r_{ss} is the number of alternate path round trips spent in slow start. Since $init\_cwnd = 1$, $b = 2$, and $r_{ss} \leq r$, we can solve for r_{ss} to arrive at

$$r_{ss} = max\left[r, 1 + log_{\frac{3}{2}}(ssthresh)\right] \tag{2.3}$$

Using a component of the slow start data transfer model from Cardwell et al. (2000), the number of TPDUs sent during the first r_{ss} round trips on the alternate path is

$$
\begin{aligned}
d_{ss} &= 1 \cdot \frac{\left(\frac{3}{2}\right)^{r_{ss}} - 1}{\frac{3}{2} - 1} \\
&= 2\left[\left(\frac{3}{2}\right)^{r_{ss}} - 1\right]
\end{aligned} \tag{2.4}
$$

The remaining round trips, r_{ca}, are the number of round trips the transfer on the alternate path spends in congestion avoidance:

$$r_{ca} = r - r_{ss} \tag{2.5}$$

During congestion avoidance, cwnd grows by 1 MTU each round trip. Thus, we use $cwnd_i$ to denote the sender's cwnd during the i-th round trip in congestion avoidance:

$$cwnd_{i+1} = cwnd_i + 1 \tag{2.6}$$

Then since a sender begins in congestion avoidance with $cwnd = ssthresh + 1$, we have:

$$cwnd_{i+1} = ssthresh + i \tag{2.7}$$

Thus, the number of data TPDUs sent during congestion avoidance is

$$
\begin{aligned}
d_{ca} &= \sum_{i=1}^{r_{ca}}(ssthresh + i) \\
&= (r_{ca} \cdot ssthresh) + \sum_{i=1}^{r_{ca}} i \\
&= (r_{ca} \cdot ssthresh) + \frac{r_{ca} + (r_{ca})^2}{2}
\end{aligned} \tag{2.8}
$$

Combining (2.4) and (2.8), we estimate the number of successful data TPDUs that PMR $= 0$ sends to the alternate destination in r alternate path round trips as

$$d = d_{ss} + d_{ca} \tag{2.9}$$

Since PMR $= 1$ only sends only one TPDU to the alternate destination in r alternate path round trips, the difference in the number of TPDUs that PMR $= 0$ and PMR $= 1$ send in this scenario is $(d - 1)$. Therefore, the relative performance difference between PMR $= 0$ and PMR $= 1$ in this scenario depends on r. When $r = 1$, it follows that $d = 1$, and thus PMR $= 0$ performs no better than PMR $= 1$. However, when $r > 1$, PMR $= 0$ outperforms PMR $= 1$ since $d > 1$.

This analysis assumes that the alternate path does not experience loss, but we now relax this constraint by considering alternate path losses after TSN 1 (the case where TSN 1 is lost is presented next in Scenario 4). Without getting into the details of such scenarios (there are an infinite number), it suffices to say that our estimate of d in (2.9) is an overestimate when loss is introduced. However, the fact that $d \geq 1$ remains true. Therefore, it remains that, in this scenario, PMR $= 0$ performs no worse than PMR $= 1$, and may outperform PMR $= 1$ by as much as $(d - 1)$ TPDUs, depending on r and the loss conditions on the alternate path.

Scenario 4

The first TPDU sent to the primary destination and the first TPDU sent to the alternate destination following TSN 1's timeout are both lost.

- **PMR = 0** TSN 1's retransmission times out first, and TSN 1 is re-retransmitted to the primary destination. When TSN 1 is acked, the failover is cancelled and normal data transfer continues to the primary destination from this point. Note that the heartbeat times out later but does not affect the data transfer.

- **PMR = 1** TSN 1's retransmission times out first, and TSN 1 is re-retransmitted to the primary destination. When TSN 1 is acked, the failover is cancelled, but the sender cannot send any new data until TSN 2 times out. Once TSN 2 times out, the sender retransmits it to the alternate destination, and sends TSN 3 to the primary destination. From this point, normal data transfer continues to the primary destination.

Similar to Scenario 3, this scenario shows that PMR $= 0$ outperforms PMR $= 1$ when the primary path experiences consecutive timeouts. Again, the improvement is based on r, but in this scenario, r is the number of primary path round trips defined as

$$r = min \left[1, \frac{RTO_1^2 - RTO_2^1}{RTT_1} \right] \qquad (2.10)$$

where RTO_1^2 is the primary path's RTO for the second consecutive timeout, RTO_2^1 is the alternate path's RTO, and RTT_1 is the primary path's RTT. Using this value of r in (2.3) and (2.5), we can use (2.9) to estimate the number of successful data TPDUs, d, that PMR $= 0$ sends to the primary destination by the time RTO_2^1 expires. Therefore, this scenario also shows PMR $= 0$ performing no worse than PMR $= 1$, and possibly outperforming PMR $= 1$ by as much as $(d - 1)$ TPDUs.

The chances of encountering each of these four scenarios depends on the loss conditions of the two paths. Regardless of which scenario is encountered when a timeout occurs on the primary path, lower PMR settings (PMR = 0 in particular) provide a transfer with more to gain (potentially several more TPDUs successfully transferred) and less to lose (at most, one less TPDU successfully transferred). Therefore, lower PMR settings do not degrade performance and may actually improve performance.

2.3.3 Permanent Failovers

When failovers are temporary, traffic migrates back to the primary path when it recovers. This migration can potentially throttle the sending rate because the sender returns to slow start's cwnd of one MTU. To avoid this slowdown, this author introduces a major potential change to SCTP – the concept of "permanent failover" using a *Change Primary Threshold (CPT)*. Permanent failover is based on a two-level threshold failover mechanism proposed in Caro et al. (2002). Once failover occurs, the sender can make the failover permanent (i.e., change the primary destination) if more than CPT heartbeat probes sent to the primary destination time out.

The specification for permanent failovers, shown in Figure 2.25, adds two new transitions to the finite state machine in Figure 2.23. While the association is in Phase II or III, if the primary destination's CPT threshold is exceeded, the primary destination is changed to the alternate destination currently in use. In Phase II, the association returns to Phase I with the new primary destination. In Phase III, however, the association remains in Phase III when a new primary destination is set; that is, changing the primary destination does not change the status of any destination, and thus the association remains in the dormant state.

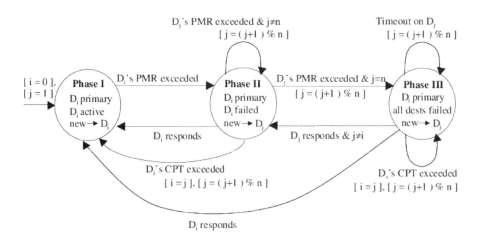

Figure 2.25 FSM for permanent failovers.

We evaluate different CPT settings using the same methodology explained in Section 2.3.2, except here we only focus on PMR = 0 and the Dormant Hop behavior. We use these settings, because we have shown in Section 2.3.2 that PMR = 0 performs best and the dormant state behavior behavior is insignificant.

Symmetric Path Delays

Figure 2.26 plots the average 80MB file transfer time for $\{3,5,8,10\}\%$ primary path loss, a 90ms primary path RTT, and a 90ms alternate path RTT. When the alternate path loss rate is lower than the primary path loss rate, more aggressive permanent failover (i.e., lower CPT settings) dramatically improve performance. On the flip side, the performance is degraded relatively little when the alternate path loss rate is higher than that of the primary path. For example, when the primary path loss rate is 5%, reducing CPT from 5 to 0 improves performance by as much as 88% and degrades performance by at most 9%.

Since paths with lower loss rates are less likely to exceed CPT, associations with lower CPT settings tend to spend less time on the higher loss rate path. The intuition is as follows. If a sender permanently fails over to a path with a higher loss rate, the performance may degrade, but only temporarily. Eventually, CPT will be exceeded again and the sender will switch back to the lower loss rate path. *Therefore, when the path delays are symmetric, the most aggressive permanent failover (i.e., CPT = 0) provides the best performance.*

Asymmetric Path Delays

Figure 2.26 shows that lowering CPT improves performance when path delays are symmetric, but what happens when path delays are asymmetric? Figure 2.27 plots the average 80MB file transfer for $\{3,5,8,10\}$ primary path loss, a 90ms primary path RTT, and a 210ms alternate path RTT. These results show that lower CPT settings *may* improve performance, but only when the alternate path's loss rate is *much* lower than the primary path loss rate. Otherwise, aggressive permanent failover degrades performance significantly. For example, when the primary path loss rate is 5%, reducing CPT from 5 to 0 improves performance 76% and 23% for 0% and 1% alternate path loss rates, respectively. On the other hand, the performance suffers (by as much as 54%) for all other alternate path loss rates. Thus, to benefit from a change primary, the difference in path delays requires an alternate path loss rate low enough to offset the alternate path's relatively large delay.

Note that worst performance for aggressive permanent failover occurs when both paths' loss rates are similar. As expected, aggressive permanent failover's performance improves as the alternate path's loss rate decreases relative to the primary's. Surprisingly, however, aggressive permanent failover's performance also improves as the alternate path's loss rate increases relative to the primary's. As explained in Section 2.3.3, lower CPT settings allows an association to reduce the time spent on the higher loss rate path. Therefore, as the alternate path's loss rate increases, the

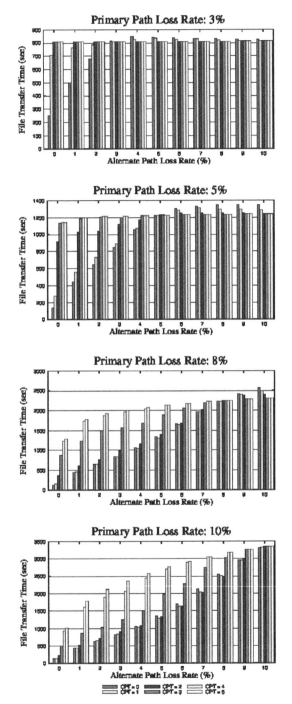

Figure 2.26 CPT evaluation: PMR = 0, 90ms primary path RTT, and 90ms alternate path RTT.

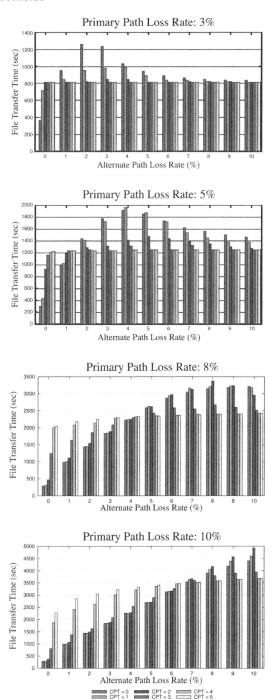

Figure 2.27 CPT evaluation: PMR $= 0$, 90ms primary path RTT, and 210ms alternate path RTT.

association will spend less time on the alternate path, thereby reducing the negative effects of its longer RTT.

Recognize that the results in Figure 2.27 present only one perspective with respect to asymmetric path delays. We show in Figure 2.28 that if the association begins with the longer delay path as the primary, lower CPT settings are advantageous regardless of the paths' loss rates. When starting on the longer delay path, the sender has only to gain with more aggressive permanent failovers. If the alternate path's loss rate is lower, the association will spend more time on the shorter delay path. Otherwise, the association will spend more time on the longer delay path, which it would have anyway with higher CPT settings.

These results seem to demonstrate that failovers should be permanent only when the alternate path has a shorter RTT, but both RTT and loss conditions need to be considered in the decision process. A path with shorter RTT and higher loss rate may provide lower throughput than a path with longer RTT and lower loss rate. With an estimated RTT and loss rate (p) for each path, a sender can apply Padhye's simplified throughput model, $\frac{1}{RTT}\sqrt{\frac{3}{4p}}$, from Padhye et al. (1998) to compare paths and determine if a permanent failover would be advantageous. Future work is to develop a mechanism to measure the loss rate of an idle alternate path without introducing unnecessary overhead.

2.3.4 Summary

We investigated the effects of reducing SCTP's failure detection threshold, Path.Max.Retrans (PMR), to less than the currently specified six consecutive timeouts. As expected, the number of spurious failovers increased as PMR was lowered, but we found that spurious failovers do not degrade performance. In fact, we found that aggressive failover settings have little effect on long term goodput averages for primary path loss rates less than 8%. At higher primary path loss rates, lower PMR settings improve goodput (even when the loss rate and/or delay is higher on the alternate path). Furthermore, since lower PMR settings provide less stalls during timeout events, short transfers may benefit even at low primary path loss rates.

We also explored the concept of permanent failovers to further improve performance by avoiding a slowdown of the sending rate after a failed path recovers. We found that that permanent failovers can improve performance if a sender has an estimate of each path's RTT and loss rate to make an informed decision.

We realize that aggressive failover thresholds may draw concern. First, traditional thinking is that frequent traffic redirection is counter-productive, but that intuition comes from research in congestion-based routing algorithms. Migrating traffic back-and-forth on an end-to-end basis does not suffer the side effects (e.g., reordering, inaccurate RTT estimates, etc.) that are introduced, for example, when an intermediate router "flip-flops" traffic between routes. These side effects are avoided because each time a flow moves to a new path, it begins from slow start as if it were a new flow. Furthermore, SCTP maintains path information (e.g., RTT, cwnd, ssthresh, etc.) per destination.

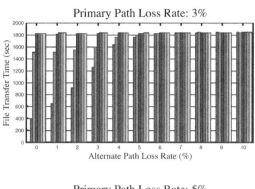

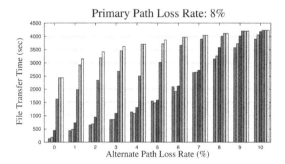

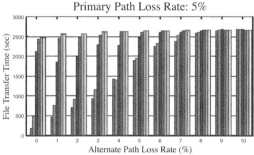

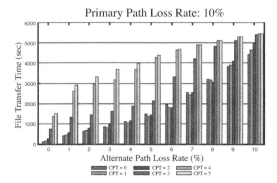

Figure 2.28 CPT evaluation: PMR $= 0$, 210ms primary path RTT, and 90ms alternate path RTT.

Second, "global failover synchronization" becomes possible with an aggressive traffic migration design. A cycle is formed when a bottleneck router drops a burst of packets, causing multiple flows to timeout and move their traffic to an alternate path. These flows then simultaneously probe their primary destination, and if successful, simultaneously migrate back to their primary path and increase their cwnds up to the point where a burst of drops occurs again.

However, we argue that global failover synchronization is no worse than the existing well-known phenomenon of global TCP congestion control synchronization (Braden et al. 1998). In both cases, synchronized timeouts cause synchronized slow starts and cwnd evolution, but in the case of failover, the cwnd evolution may occur on alternate paths that do not share bottlenecks. If so, a single flow's traffic migration appears no different than a new end-to-end flow, because each time a flow migrates to a new path, the flow begins from slow start with a cwnd of one MTU. In fact, since new flows may begin with a cwnd as large as four MTUs (Allman et al. 2002), a single flow's traffic migration is more conservative than a new flow.

On the other hand, if multiple flows do migrate to alternate paths that share a bottleneck, these flows will not disturb the network any more than a synchronized TCP timeout would. In both cases, multiple flows begin from slow start with cwnd = one MTU, and simultaneously grow their cwnd. The only difference being that in the case of failover, the cwnd evolution happens to be on a different path than where the synchronized timeout occurred. In any case, AQM techniques eliminate global synchronization (Braden et al. 1998), which also includes global failover synchronization.

2.4   Related Work and Conclusions

We conclude this chapter with a summary of related work (Section 2.4.1), a summary of our key results (Section 2.4.2), and suggestions for future study (Section 2.4.3).

2.4.1   Related Work

We present work that aims to improve path failure resilience without using transport layer multihoming. These approaches include dynamic host routing, site multihoming, server replication, overlay routing networks, and interrupted connection re-establishment.

Dynamic Host Routing

Dynamic host routing is an approach used by Touch and Faber to address network fault tolerance at the end hosts using the routing layer (Touch and Faber 1997). Their approach strives towards minimal changes, including no operating system modifications and no changes to the application or transport layers. Their solution combines the use of multihomed hosts, virtual network interfaces, and the RIP (Hedrick 1988) routing protocol running on end hosts. Each multihomed host is configured with a virtual network interface that connects to the host's real network interfaces through virtual links. All connections to and from the multihomed host bind to the virtual interface IP address, and RIP is used to dynamically route traffic through the appropriate real interface. This approach allows new and existing transport layer connections to be resilient to network failures. Since RIP needs to run on all hosts that wish to communicate resiliently, dynamic host routing is intended for small scale networks and does not scale well for the Internet.

Transport layer multihoming offers better scalability, because a host only maintains multipath reachability information for peers to which it is connected. Another advantage offered by transport layer multihoming is that actual data transfers serve as implicit probes; hence, the transport layer can detect failure more quickly than a routing protocol (assuming that the application sends data more frequently than a routing protocol sends probes).

Site Multihoming

Site multihoming is when an entire site has more than one connection to the Internet. Those connections can be either through the same ISP or different ISPs. Today, multihomed sites obtain a dedicated block of address space, and advertise its prefix route through each of its ISP connections. Often the route for the prefix propagates to all routers connected to the fault-free zone, which increasingly burdens BGP's convergence time. Routing stability is threatened as this type of site multihoming increases.

The IETF's *multi6* working group seeks alternative approaches that offer more scalability, but their work focuses on IPv6 and requires BGP support (Abley et al. 2003). Some commercial products (Avaya: Adaptive Networking Software (Route-Science) 2004; Fat Pipe Networks 1989; Internap: Network-based Route Optimization 1996; Radware 1997; Stonesoft: StoneGate Multi-link Technology 1990) offer site multihoming capabilities without BGP support. In any case, site multihoming protects against a single access link failure, but cannot avoid the long convergence times of BGP within the Internet.

Server Replication

Server replication aims to avoid failed network paths and overloaded servers by redirecting client requests to a redundant server. Content Distribution Networks (CDNs), such as (Akamai 1998) and Coral (Freedman et al. 2004), cache content at intermediate nodes scattered around the Internet. Client requests are then redirected using DNS or server redirects. CDNs are effective at maintaining availability during flash crowds, but they focus on replicating content instead of improving host availability. Thus, CDNs require additional reliability mechanisms to protect uncached content from failures.

Server pools avoid this problem by replicating complete servers in clusters. Several projects exist which address different aspects of the goal. The IETF's *rserpool* working group is developing an architecture and protocols for the management and operation of server pools (Tuexen et al. 2002). The TCP Connection Migration (Snoeren et al. 2001) and Migratory TCP (M-TCP) (Sultan et al. 2002) projects provide mechanisms for migrating and resuming a service session from an overloaded server to a redundant backup server. A server is considered overloaded if its system resources or network resources are fully utilized, but these projects do not address the issues in determining when a server is overloaded.

SPAND maintains a shared repository of passive statistics of Internet transfers that can be used for future transfers to choose a good server from a pool (Seshan et al. 1997). Smart Clients use downloaded portable code to handle server selection at the client side for better flexibility and performance (Yoshikawa et al. 1997). Crovella et al. show that latency-based server selection outperforms random selection and techniques based on geographic location or network hop count (Crovella and Carter 1995). Dykes et al. independently also conclude in their study of server selection techniques that a latency-based approach is best (Dykes et al. 2000). They also conclude that picking the first server to respond to a TCP SYN packet is more effective than using statistical records of previous transfers.

Research and actual deployment of server replication techniques have shown them to provide extra resilience against path failures. Unfortunately, this approach is expensive and only high-end Web sites can afford to add server redundancy. Moreover, this approach does not improve host availability for hosts that act as clients and do not have content to serve.

Overlay Routing Networks

A Resilient Overlay Network (RON) is an overlay architecture that allows end-to-end communication over the Internet to detect and recover from path outages (Andersen et al. 2001). A RON consists of a group of nodes which together form an application layer overlay on top of the existing Internet routing infrastructure. The RON nodes monitor the Internet paths among themselves to decide between routing packets directly over the Internet or via another RON node. These decisions are optimized for the individual needs of applications running on the RON nodes.

NATRON (Yip 2002) extends RON by using a Network Address Translator (NAT) (Egevang and Francis 1994) to reach external hosts, similar to a NAT-based proposal by the Detour project (Collins 1998). Packets that are routed through the NATRON infrastructure are tunneled to an intermediate NATRON node. The intermediate nodes use NAT to replace the source address with its own, and forwards packets to their target destination. The masqueraded IP address cause reply packets to be sent to the same intermediate node, giving NATRON the ability to control routing in both directions.

SOSR (Scalable One-hop Source Routing) is another NAT-based solution, but the decision making is done by the source node for scalability (Gummadi et al. 2004). When a source detects a failure, it selects one or more intermediate nodes to route through. The source simply configures a SOSR node as its IP gateway, and its packets are automatically routed through the overlay to the destination node. Intermediate nodes forwards messages it receives to the destination, acting as a NAT proxy for the source. SOSR is transparent to the destination.

Multihomed Overlay Networks (MONET) builds on the RON concept by combining an overlay network with multihoming to address path failures both at access links and within the Internet (Andersen 2005). Each MONET node is a proxy that serves as a conduit for client connections to a server. MONETs offer more scalability than RONs, because communication is not confined to a small group. MONETs are able to increase availability of existing unmodified servers.

Overlay approaches work well to route around failures, but they have some limitations. They require extra infrastructure that is not scalable due to per flow state within the infrastructure. The RON-based approaches monitor path quality to select the best available path, but this required background traffic is also not scalable beyond a small set of nodes. NAT-based solutions inherit all the issues with NAT (which are many). Furthermore, ISPs may attempt to block overlay approaches, because they do not respect BGP routing policies.

Interrupted Connection Re-establishment

Sometimes failures are unavoidable and transport layer connections are interrupted. Several projects, such as Reliable Sockets (Rocks) (Zandy and Miller 2002), MSOCKS (Maltz and Bhagwat 1998), and TCP Migrate (Snoeren 2002; Snoeren and Balakrishnan 2000), focus on providing session continuity after a disconnection or change of network attachment (i.e., mobility). However, these techniques do not attempt to maintain end-to-end connectivity in the presence of path failures.

The Mobile SCTP (Riegel and Tuexen 2004) project exploits the Dynamic Address Reconfiguration extension (Stewart et al. 2005b) for SCTP to provide yet another mechanism for transport layer mobility. This technique differs in that it aims to actually maintain end-to-end connectivity instead of disconnection restoration.

2.4.2 Conclusions

Transport layer multihoming aims to provide added fault tolerance needed for persistent sessions, i.e., maintain end-to-end connectivity in the presence of failures. Although connectivity may be maintained, a transfer may suffer from periods of inactivity due to conservative use of alternate paths. We have investigated retransmission policies and failover mechanisms that attempt to minimize stall time during network congestion and failure events.

The decision tree in Figure 2.29 summarizes all the results in this chapter by suggesting a retransmission policy and failover threshold based on expected network conditions. Recognize that we are using simplistic simulation models and discrete network parameters to suggest exact values for real, complex network conditions, so our results/suggestions are only meant to provide intuition. Left branches require stricter criteria from network expectations, but are able to suggest aggressive use of alternate paths. Right branches suggest less aggressive use of alternate paths due to their looser network requirements.

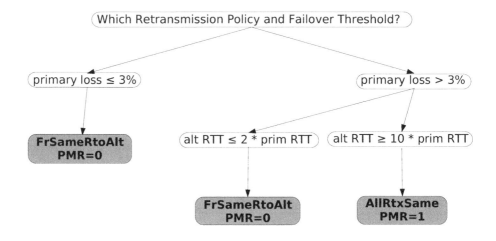

Figure 2.29 Summary of results.

The Internet provides an environment where an endpoint can be more aggressive in using alternate paths. Since most paths on the Internet experience less than 3% loss (Zhang et al. 2001), FrSameRtoAlt with PMR = 0 should provide the best performance. Even if we expect higher loss rates on the Internet, path delays generally are *roughly* symmetric (i.e., no path delay is more than double of another) and proactive routing is used. Therefore, FrSameRtoAlt with PMR = 0 is still a good choice for the Internet.

If absolutely nothing is known or expected of a network, then the most conservative settings should be used. Following the rightmost branches of the decision tree in Figure 2.29, these settings are AllRtxSame and PMR = 1. With these "conservative" settings, an end-to-end failover takes only two consecutive timeouts (i.e., 3 seconds with RFC2960's recommended minimum RTO setting). Note, however, that these settings are significantly more aggressive than those recommended currently for SCTP in RFC2960, but are conservative enough to perform well in topologies with high loss rates and highly asymmetric path delays.

Regarding our introduction and investigation of permanent failover, permanent failovers can improve performance when the sender has an accurate estimate of the RTT and loss rate of each path before making a retransmission decision. Otherwise, we show that permanent failover is not recommended; failovers should be temporary, and traffic should resume on the primary path when it recovers.

In essence, we have found that traditional conservative failover techniques used in routing do not apply when path redundancy begins at the end hosts and is handled by the transport layer. Since failovers at the routing layer are transparent to the transport layer, the failover thresholds must be conservative to avoid oscillations that could cause the transport layer to maintain inaccurate path metrics (RTT, cwnd, ssthresh). On the other hand, a multihomed transport layer is completely aware of failover events and is able to maintain separate metrics per path. As a result, transport layer multihoming can improve performance by providing aggressive failovers that reduce stall time during network congestion and failure events. These results provide insight on design decisions for future multihomed transport protocols.

2.4.3 Future Work

Parallel initiation requests is a future work area worth considering. Although transport layer multihoming provides failure resilience to established associations, it does not provide a mechanism for establishing associations in the face of path failure. DNS returns a list of IP addresses for multihomed hosts, and applications are responsible for attempting more than one address until success is obtained Braden (1989a). Current applications pass only one peer IP address to a multihomed transport layer, and the transport layer then learns the other peer addresses from the peer during association establishment. However, if the peer does not respond, the initiating transport layer has no other peer addresses to try, and can only retry the same peer address. After several failed attempts, the initiating transport layer gives up, and the application is forced to try another peer IP address.

It would be ideal if applications could simply pass the list of IP addresses to a multihomed transport protocol, and handover the responsibility of trying multiple peer addresses, if necessary, during transport layer association initiation. Part of the problem with passing a list to the transport layer is that an application cannot assume that the list of IP addresses returned by DNS refers to a single host, because DNS also returns a list for replicated servers (e.g., Akamai Akamai (1998)). This ambiguity poses a problem if a transport layer relies on an application passing a list of addresses, because the transport layer may potentially try to establish an association consisting of IP addresses for physically different hosts.

We propose a mechanism for future work that allows an association to be established during path failures. This mechanism requires that multihomed transport protocols accept a list of IP addresses from applications, but not assume that the entire list can be bound to a single association. Instead, the transport layer assumes, for all practical purposes, that all the addresses refer to different physical hosts. This way, the transport layer can try multiple addresses during association initiation if necessary. Once a peer replies, the normal establishment procedure occurs, and the initiating transport layer binds the association to the address list reported by the peer (and not the one passed by the application).

We also propose an enhanced version of this mechanism that can be used for determining the path (or redundant server) with lowest latency, which can be used for selecting the primary path. When initiating an association, a multihomed transport layer sends multiple initiation requests in parallel – one to each address in the list provided by the application. The first acked request is used to establish an association, and the others are ignored. Again, the initiating transport layer binds the association to the address list reported by the peer. This method of selecting a primary path (or redundant server) is similar to Dykes et al.'s suggestion for selecting a server from a cluster of replicas (Dykes et al. 2000).

Bibliography

Akamai 1998. www.akamai.com.

Abley J, Black B and Gill V 2003 Goals for IPv6 Site-Multihoming Architectures. RFC3582, IETF.

Allman M, Floyd S and Partridge C 2002 Increasing TCP's Initial Window. RFC3390, IETF.

Allman M, Paxson V and Stevens W 1999 TCP Congestion Control. RFC2581, IETF.

Andersen D 2005 *Improving End-to-End Availability Using Overlay Networks* PhD thesis Dept of Electrical Engineering and Computer Science, MIT.

Andersen D, Balakrishnan H and M. Kaashoek RM 2001 Resilient Overlay Networks *18th ACM Symposium on Operating Systems Principles (SOSP 2001)*, Banff, Canada.

Avaya: Adaptive Networking Software (RouteScience) 2004. www.avaya.com/gcm/master-usa/en-us/products/offers/adaptive_networking_software.htm.

Berkeley U, LBL, USC/ISI and Parc X 2003 ns-2 Documentation and Software. www.isi.edu/nsnam/ns.

Braden B, Clark D, Crowcroft J, Davie B, Deering S, Estrin D, Floyd S, Jacobson V, Minshall G, Partridge C, Peterson L, Ramakrishnan K, Shenker S, Wroclawski J and Zhang L 1998 Recommendations on Queue Management and Congestion Avoidance in the Internet. RFC2309, IETF.

Braden R 1989a Requirements for Internet Hosts – Application and Support. RFC1123, IETF.

Braden R 1989b Requirements for Internet Hosts – Communication Layers. RFC1122, IETF.

CAIDA: Packet Sizes and Sequencing 1998. traffic.caida.org.

Cardwell N, Savage S and Anderson T 2000 Modeling TCP Latency *IEEE INFOCOM 2000*, Tel Aviv, Israel.

Caro A and Iyengar J 2006 ns-2 SCTP module. http://pel.cis.udel.edu.

Caro A, Iyengar J, Amer P, Heinz G and Stewart R 2002 A Two-level Threshold Recovery Mechanism for SCTP *SCI 2002*, Orlando, FL.

Caro A, Iyengar J, Amer P, Ladha S, Heinz G and Shah K 2003 SCTP: A Proposed Standard for Robust Internet Data Transport. *IEEE Computer* **36**(11), 56–63.

Chandra B, Dahlin M, Gao L and Nayate A 2001 End-to-End WAN Service Availability *3rd USENIX Symposium on Internet Technologies and Systems (USITS)*, San Francisco, CA.

Claffy K, Miller G and Thompson K 1998 The Nature of the Beast: Recent Traffic Measurements from an Internet Backbone. *INET 1998*.

Collins A 1998 *The Detour Framework for Packet Rerouting*. MS Thesis, Dept of Computer Science and Engineering, University of Washington.

Crovella M and Carter R 1995 Dynamic Server Selection in the Internet *Third IEEE Workshop on the Architecture and Implementation of High Performance Communication Subsystems (HPCS)*.

Duke M, Henderson T, Spagnolo P, Kim J and Michael G 2003 Stream Control Transport Protocol (SCTP) Performance Over the Land Mobile Satellite Channel *MILCOM 2003*, Boston, MA.

Dykes S, Robbins K and Jeffery C 2000 An Empirical Evaluation of Client-Side Server Selection Algorithms *IEEE INFOCOM 2000*, Tel Aviv, Israel.

Egevang K and Francis P 1994 The IP Network Address Translator (NAT). RFC1631, IETF.

Fat Pipe Networks 1989. www.fatepipeinc.com.

Freedman M, Freudenthal E and Mazieres D 2004 Democratizing Content Publication with Coral *First Symposium on Networked Systems Design and Implementation (NSDI)*, San Francisco, CA.

Gummadi K, Madhyastha H, Gribble S, Levy H and Wetherall D 2004 Improving the Reliability of Internet Paths with One-Hop Source Routing *6th USENIX OSDI*, San Francisco, CA.

Gurtov A, Passoja M, Aalto O and Raitola M 2002 Multi-Layer Protocol Tracing in a GPRS Network *International Conference on Ubiquitous Computing*.

Hedrick C 1988 Routing Information Protocol. RFC1058, IETF.

Inamura H, Montenegro G, Ludwig R, Gurtov A and Khafizov F 2003 TCP over Second (2.5G) and Third (3G) Generation Wireless Networks. RFC3481.

Internap: Network-based Route Optimization 1996. www.internap.com/products/networkbased.htm.

Iren S, Amer P and Conrad P 1999 The Transport Layer: Tutorial and Survey. *ACM Computing Surveys*.

Iyengar J, Amer P and Stewart R 2004a Concurrent Multipath Transfer Using Transport Layer Multihoming: Performance Under Varying Bandwidth Proportions *MILCOM 2004*, Monterey, CA.

Iyengar J, Amer P and Stewart R 2004b Retransmission Policies for Concurrent Multipath Transfer Using SCTP Multihoming *ICON 2004*, Singapore.

Iyengar J, Shah K, Amer P and Stewart R 2004c Concurrent Multipath Transfer Using SCTP Multihoming *SPECTS 2004*, San Jose, California.

Jayaram R and Rhee I 2003 A Case for Delay-based Congestion Control for CDMA 2.5G Networks *International Conference on Ubiquitous Computing*.

Karn P and Partridge C 1987 Improving Round-Trip Time Estimates in Reliable Transport Protocols *ACM SIGCOMM 1987*.

Kohler E, Handley M and Floyd S 2004 Datagram Congestion Control Protocol (DCCP). draft-ietf-dccp-spec-09.txt. (work in progress).

Labovitz C, Abuja A and Jahanian F 1999 Experimental Study of Internet Stability and Wide-Area Backbone Failures *29th International Symposium on Fault-Tolerant Computing (FTCS 1999)*, Madison, WI.

Labovitz C, Abuja A, Bose A and Jahanian F 2000 Delayed Internet Routing Convergence *ACM SIGCOMM 2000*, Stockholm, Sweden.

Ladha S, Baucke S, Ludwig R and Amer P 2004 On Making SCTP Robust to Spurious Retransmissions. *ACM SIGCOMM Computer Communication Review* **34**(2), 123–135.

Leland W, Taqqu M, Willinger W and Wilson D 1993 On the Self-similar Nature of Ethernet Traffic *ACM SIGCOMM 1993*, San Francisco, CA.

Ludwig R and Katz R 2000 The Eifel Algorithm: Making TCP Robust Against Spurious Retransmissions. *ACM Computer Communications Review* **30**(21), 30–36.

Maltz D and Bhagwat P 1998 MSOCKS: An Architecture for Transport Layer Mobility *IEEE INFOCOM 1998*, San Francisco, CA.

Moore D, Voelker G and Savage S 2001 Inferring Internet Denial of Service Activity *2001 USENIX Security Symposium*, Washington, D.C.

Padhye J, Firoiu V, Towsley D and Kurose J 1998 Modeling TCP Throughput: A Simple Model and its Empirical Validation *ACM SIGCOMM 1998*, pp. 303–314, Vancouver, CA.

Paxson V 1997 End-to-End Routing Behavior in the Internet. *IEEE/ACM Transactions on Networking* **5**(5), 601–615.

Radware 1997. www.radware.com.

Rekhter Y and Li T 1995 A Border Gateway Protocol 4 (BGP-4). RFC1771, IETF.

Rekhter Y, Li T and Hares S 2002 A Border Gateway Protocol 4 (BGP-4). draft-ietf-idr-bgp4-18.txt, Internet Draft (work in progress), IETF.

Riegel M and Tuexen M 2004 Mobile SCTP. draft-riegel-tuexen-mobile-sctp-05.txt (work in progress), IETF.

Seshan S, Stemm M and Katz R 1997 SPAND: Shared Passive Network Performance Discovery *1st USENIX Symposium on Internet Technologies and Systems (USITS)*, Monterey, CA.

Snoeren A 2002 Enabling Internet Suspend/Resume with Session Continuations *Student Oxygen Workshop 2002*.

Snoeren A and Balakrishnan H 2000 An End-to-End Approach to Host Mobility *ACM MobiCom 2000*, Boston, MA.

Snoeren A, Andersen D and Balakrishnan H 2001 Fine-Grained Failover Using Connection Migration *3rd USENIX Symposium on Internet Technologies and Systems (USITS)*, San Francisco, CA.

Stewart R and Xie Q 2001 *Stream Control Transmission Protocol (SCTP): A Reference Guide*. Addison Wesley, New York, NY.

Stewart R, Arias-Rodriguez I, Poon K, Caro A and Tuexen M 2005a Stream Control Transmission Protocol (SCTP) Implementer's Guide. draft-ietf-tsvwg-sctpimpguide-13.txt, Internet Draft (work in progress), IETF.

Stewart R, Ramalho M, Xie Q, Tuexen M and Conrad P 2005b Stream Control Transmission Protocol (SCTP) Dynamic Address Reconfiguration. draft-ietf-tsvwg-addip-sctp-11.txt, Internet Draft (work in progress), IETF.

Stewart R, Xie Q, Morneault K, Sharp C, Schwarzbauer H, Taylor T, Rytina I, Kalla M, Zhang L and Paxson V 2000 Stream Control Transmission Protocol. RFC2960.

Stonesoft: StoneGate Multi-link Technology 1990. www.stonesoft.com/products/ISP_Multi-homing.

Strayer T and Weaver A 1988 Evaluation of Transport Protocols for Real-time Communications. Technical Report TR-88-18, CS Dep., University of Virginia.

Sultan F, Srinivasan K, Iyer D and Iftode L 2002 Migratory TCP: Highly Available Internet Services using Connection Migration *22nd International Conference on Distributed Computing Systems (ICDCS)*, Vienna, Austria.

Touch J and Faber T 1997 Dynamic Host Routing for Production Use of Developmental Networks *ICNP 1997*, Atlanta, GA.

Tuexen M, Xie Q, Stewart R, Shore M, Ong L, Loughney J, Loughney J and Stillman M 2002 Requirements for Reliable Server Pooling. RFC3237, IETF.

Walrand J 1991 *Communication Networks: A First Course*. Aksen Associates.

Willinger W, Taqqu M, Sherman R and Wilson D 1995 Self-Similarity Through High-Variability: Statical Analysis of Ethernet LAN Traffic at the Source Level *ACM SIGCOMM 1995*, Cambridge, MA.

Yip A 2002 *NATRON: Overlay Routing to Oblivious Destinations*. MS Thesis, Dept of Electrical Engineering and Computer Science, MIT.

Yoshikawa C, Chun B, Eastham P, Vahdat A, Anderson T and Culler D 1997 Using Smart Clients to Build Scalable Services *USENIX Annual Technical Conference*, Anaheim, CA.

Zandy V and Miller B 2002 Reliable Network Connections *ACM MobiCom 2002*, Atlanta, GA.

Zhang Y, Paxson V and Shenker S 2001 On the Constancy of Internet Path Properties *CM SIGCOMM Internet Measurement Workshop (IMW 2001)*, San Francisco, CA.

3

Support of Node Mobility between Networks

F. Richard Yu, Li Ma and Victor C.M. Leung

The available of multiple radio interfaces in mobile nodes provides the means and incentive to apply multihoming to support node mobility between networks. Stream Control Transmission Protocol (SCTP) with the Dynamic Address Reconfiguration extension provides a good solution to support node mobility or handovers between networks. Compared to the techniques based on network layer Mobile Internet Protocol (MIP) and application layer Session initiation Protocol (SIP), the

SCTP-based vertical handover scheme has many advantages, including simpler network architecture, improved throughput and delay performance. Seamless handovers that minimizes the disruption of the end-to-end connection is particularly important for multimedia communications. In this chapter, we give an overview of the node mobility problem and mobile SCTP (mSCTP) as a general solution of this problem. We then describe how we apply mSCTP to multiple wireless interfaces supporting different wireless access technologies to support seamless vertical handover support across heterogeneous wireless networks. We also present an error recovery scheme to improve SCTP performance during vertical handover when packet losses occur in wireless links. Simulations results are presented to demonstrate the effectiveness of the proposed schemes.

3.1   Introduction

Recently, there has been significant growth in the use of wireless services. There exist disparate wireless systems, such as Bluetooth (Johansson et al. 2001) and ultra-wideband (UWB) radios (Yang and Giannakis 2004) for personal areas, wireless local area networks (WLANs) for local areas, WiMAX for metropolitan areas, 2G/3G/4G cellular networks for wide areas, and satellite networks for global networking. With the proliferation of wireless networking standards, increasingly portable electronic devices are being equipped with multiple transceivers to enable them to access different wireless networks. For example, 3G cellular smart phones are available with built-in WLAN and Bluetooth interfaces. The complementary characteristics of different wireless networks make it attractive to integrate these heterogeneous wireless access technologies. The next generation wireless communication systems aim at integrating different wireless access technologies together and providing user always best connected (ABC) to mobile users (Gustafsson and Jonsson 2003), (Murata et al. 2009). Users would benefit from the lower overall cost and the enhanced performance of the integrated services.

Next generation heterogeneous wireless networks present significant technical challenges that require advancements in wireless networks architecture, protocol, and management algorithms. For example, while cellular networks provide always on, wide area connectivity with relatively low data rates to users with high mobility, WLANs offer much higher data rates to users with low mobility over small areas. Efficient mobility management scheme is crucial in this cellular/WLAN integration. Mobility management solution consists of support for roaming, which provides the reachability of mobile users, and support for handover (also referred as handoff in the literature), which provides ongoing connection continuity, in spite of movements across and between cellular and WLANs. Handover between heterogeneous networks are commonly referred as *vertical handovers*, whereas handovers taking place in homogeneous networks are called *horizontal handovers* (Salkintzis 2004). Horizontal handover management is normally implemented at data link layer since homogeneous underlying technologies belong to the same administrative domain. However, it is difficult to apply this approach to vertical handover where the underlying technologies are heterogeneous and usually administrative-separated. Therefore, vertical handover management should operate at a higher layer in order to provide uniform management.

Several schemes have been proposed to solve the vertical handover problem in heterogeneous wireless networks. Their operation scope varies from network to application layer. Mobile IP (MIP) (Perkins 2002) from the Internet Engineering Task Force (IETF) is a network layer solution. By inserting a level of indirection into the routing architecture, MIP provides transparent support for host mobility, including the maintenance of active Transmission Control Protocol (TCP) connections and User Datagram Protocol (UDP) port bindings. In this scheme, a home agent and a foreign agent are used to bind the home address of a mobile host to the care-of-address at the visited network and provide packet forwarding when the MH is moving between IP subnets. Triangular routing of all incoming packets to the mobile host via the home network can cause additional delays and waste of bandwidth capacity. If the correspondent host has the knowledge of where the mobile host is located, it can send packets directly to the care-of-address of the mobile host, thus enabling route optimization. The Session Initiation Protocol (SIP) (Handley et al. 1999) based approach (Wu et al. 2005) aims to keep mobility support independent of the underlying wireless access technologies and network layer elements. SIP is an application layer protocol. When a mobile host moves during an active session into a different network, first it receives a new network address, and then sends a new session invitation to the correspondent host. Subsequent data packets are forwarded to the mobile host using this new address.

Although both MIP- and SIP-based approaches can provide some level of vertical handover support in heterogeneous wireless networks, experiments have shown that it is difficult to maintain the continuity of ongoing data sessions during the handover, due to the long handover latency (Ma et al. 2004). Mobile users may experience quality of service (QoS) degradation or session disruption/termination during vertical handovers if these approaches are used. In this chapter, we present a transport-layer scheme to support vertical handovers in heterogeneous wireless networks (Ma et al. 2004). Unlike techniques based on MIP or SIP, this approach follows the end-to-end principle (Saltzer et al. 1984) in the Internet: anything that can be done in the end system should be done there. Since the transport layer is the lowest end-to-end layer in the Internet protocol stack, it is a natural candidate for vertical handover support. Moreover, in the transport-layer approach, no third party other than the endpoints participates in vertical handover, and no modification or addition of network components is required, which makes this approach universally applicable to all present and future network architectures.

In addition, we show that SCTP performance is affected by packet losses over wireless networks due to noise and interference. We then present an improvement to SCTP called Sending buffer Multicast-Aided Retransmission with Fast Retransmission (SMART-FRX) (Ma et al. 2007). Some simulation results are presented to the effectiveness of the proposed schemes.

The rest of the chapter is organized as follows: Section 2 presents the vertical handover problem and some approaches. The SCTP-based vertical handover scheme is described in Section 3. Section 4 presents the SMART-FRX scheme. Numerical and simulation results are given in Section 5. Finally, we conclude the chapter in Section 6.

3.2   Node Mobility Support in Wireless Networks

In this section, we present the integrated heterogeneous wireless network. We then introduce the approaches to support vertical handover at different layers.

3.2.1   Integrated Heterogeneous Wireless Networks

Generally speaking, there are two different ways of integrating heterogeneous wireless networks, defined as tight coupling and loose coupling interworking. Figs. 3.1 and 3.2 show the architectures for the integration. In these figures, we have two different networks, Network 1 and Network 2. A 3G cellular network is used as an example of Network 1, and a WLAN can be an example of Network 2. In reality, the number of heterogeneous networks is not limited to two.

In a tightly coupled system, as shown in Figure 3.1, a network is connected to another network in the same manner as other radio access networks. Note that many wireless networks, such as Universal Mobile Telecommunication System (UMTS), have the interfaces to other radio access networks. In this approach, these two networks could use the same authentication, mobility, and billing infrastructure, and the gateway of one network needs to implement the protocols required in the other network. The main advantage of this approach is that the existing mechanisms for authentication, mobility and QoS can be reused directly over another network. However, this approach requires modifications of the design of network 1 to accommodate the increased traffic from network 2.

In a loosely coupled system, as shown in Figure 3.2, the heterogeneous wireless networks are not connected directly. Instead, they are connected to the Internet. In this approach, different mechanisms and protocols are used to handle authentication, mobility and billing, and the traffic from one network would not go through the other network. Nevertheless, as peer IP domains, they can share the same subscriber database for functions such as security, billing and customer management.

Since wireless mobile users are free to move in integrated heterogeneous wireless networks, the support of handover between these networks, which provides ongoing service continuity and seamlessness, is needed in this integration. Handover in a heterogeneous network environment is different from that in a homogeneous environment, where it occurs only when a mobile user moves from one base station (or access point) to another. While handover within a homogeneous system is called horizontal handover, handover between different wireless access technologies is referred to as vertical handover. There are several networking protocols at different layers that enable the integration of these heterogeneous wireless networks. We present them below.

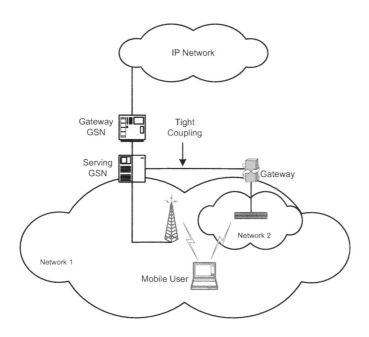

Figure 3.1 Integrated heterogeneous networks (tightly coupled).

3.2.2   Network Layer Approaches

As the identification of terminals at the network layer, a terminal's IP address highly relates to the network that the terminal attaches on. When terminals roams between different networks, the obtained IP address cannot be the same. To solve this problem, IETF proposes a Mobile IP (MIP) protocol to keep the IP consistence by adding some transferring nodes in the network (Perkins 2002).

There are totally two agents deployed in each subnets, namely home agent (HA) and foreign agent (FA). HA stores the registration information of all terminals originated from the current network. When a mobile node (MN) roams into a foreign network, it will firstly obtain a care of address (CoA) from the FA in the new network, where the CoA is usually a special IP address broadcasted by the FA. MN will further notify the HA in his home network of the CoA through a registration message. To reduce the number of registration messages, MIP is further extended with paging functionality to reduce the signaling cost in address update (Haverinen and Malinen 2001). Whenever receiving the registration message, HA will forward all incoming packets from corresponding node (CN) to FA by encapsulating them with the CoA as the destination address. And FA will thereby decapsulate the tunneled packets and send them to the MN. Whereas the outgoing packets from the MN is sent out to

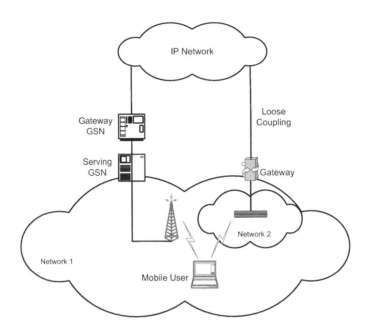

Figure 3.2 Integrated heterogeneous networks (loosely coupled).

the CN as normal. Such asymmetric routing mechanism leads to a "triangle routing" problem, where the un-optimal delivery of the incoming packets will incur heavy data transmission cost along with long packet delivery time. In Perkins and Johnson 2001, CN will cache the CoA of the MN and directly tunnel the packets to MN. The cache will be updated when receiving update messages from the HA of the home network. Note that due to the security issue, the CN only trust the notification messages sent from HA, which has a fixed authenticated IP address. Another problem here is that the handover is not seamless, where MN does not receive any packets until the handover completes and packets sent out during this period are either lost or deferred. To address this problem, Perkins and Johnson 2001 also propose to set up a temporary connections between the FAs of the previous network (oFA) and the new network (nFA). Once MN registers in the new network, it will tell its new CoA to the oFA before sending out the registration message and the oFA will forward all incoming packets to the nFA until the handover is finished. Since the oFA is usually closer to nFA than HA and CN, then packets interruption period will be greatly shortened. To further shorten the interruption period, Hsieh et al. 2003 also proposes to predict the potential handovers by tracking the MN's locations. Whenever MN is going to move out of the WLAN coverage area, it will initiate the handover in advance so that the handover completes before the WLAN signals become undetectable.

In MIP, MN should register to the HA every time when it changes the IP address. If MN moves frequently between different subnets of a foreign network, it will send back the registration messages frequently to the HA, which will generate heavy signaling traffic as well as long handover delay if HA is far away from MN. A series of schemes are proposed to manage such kind of "intra-domain" handover by localizing the address update within the foreign network. According to their functioning manner, these solutions can be classified into tunnel-based management (Gustaffson et al. 2001) (Misra et al. 2002) and routing-based management (Soliman et al. 2004) (Perkins and Wang 1999). In tunnel-based schemes, a lot of registration agents distribute hierarchically in the foreign network and data packets are tunneled between the agents at different layers. Whenever MN roams between subnets, it will register the new address to the nearest common agents instead of update in the home agent. In routing-based schemes, the MN will keep the same IP address by modifying routing tables of routers.

Generally, network layer solutions can provide universal mobility management over IP networks. However, it has two major drawbacks. It requires a lot of changes on networks, such as the set up of HA and FA in all networks. It also requires the modification of CN to implement the routing optimization. Moreover, it requires to assign a static IP address to each MN, which is very costly in the IPv4 network that short of IP addresses. Therefore, better mobility management solutions are needed.

3.2.3 Application Layer Approaches

As a signaling protocol in application layer, SIP has been widely adopted by IETF and 3GPP to help establish sessions in packet-switched networks (Handley et al. 1999). The session is multiple data communications among multiple participants. SIP also affords various mobility supports and allows a connection to be set up in the middle of a session. It is generally envisioned that the SIP should be the main signaling protocols in next generation wireless networks. In SIP, the MN is identified with a unique logical SIP address in the format of email address. When the MN roams into the foreign domain, it obtains a contact address with the domain name in the foreign domain. The MN also registers the contact address to the registration server in the home domain through the SIP proxy. CN sends out an "INVITE" message via its SIP proxy to home network of MN. The address of the home network can be figured out from the public SIP address of the MN. This request is captured by the SIP proxy in the home network of MN. After checking the registration server, the proxy in the home network of MN knows that currently MN in a foreign network. Then, the SIP proxy will return a "moved temporarily" message to CN with the current contact address of MN attached. SIP proxy of CN re-sends the "INVITE" message to MN directly. And MN returns a "OK" message and negotiate the session parameters. CN will send an "ACK" message to confirm it accepts the modification. When the negotiation is done, the session with several connections will be directly set up between CN and MN *B*. If the MN roams to a new domain in the middle of the session, it will notify the CN of its new contact address and the updated session description by another "INVITE" message. The CN further re-sets up the connection to the new location.

In comparison with the mobility solutions at other layers, SIP can be easily deployed in the network without requiring the modification of network entities or end-user protocol stacks. Moreover, SIP can be easily extended or escalated due to the operation at the highest layer and the use of text-based signaling messages. Having been accepted by a standard protocol in both Internet and tele-network, SIP is regarded as an attractive candidate to support mobility in next generation networks.

Apart from all those advantages, SIP also suffers from the long handover delay due to its operation at the highest layer. This handover delay has been proved to be unacceptable for supporting real-time multimedia services (Wu et al. 2005) and many schemes are also proposed to deal with this problem. A hierarchical mobile SIP (HMSIP) is proposed in Vali et al. 2003, where a global network proxy is set up for each domain to manage the roaming inside the domain. When MN roams between different sub-nets of the domain, it will sent out the registration messages to the global network proxy instead of home proxy. And the global network proxy will check all incoming packets and redirect them to the current address of the MN. In this way, the signaling messages for intra-domain roaming is localized in the foreign network, thereby the handover delay will be greatly shortened.

Above schemes only solve the handover between sub-nets of the same domain, which can cooperate easily due to the the homogeneous underlying technologies. However, handovers between different domains are more difficult to address for the following reasons. First, it is difficult to localize signaling messages of inter-domain handovers since it is commercially impractical to set up intermediate node between heterogeneous networks. Second, the above schemes reduce the handover delay by shortening the registration period. However, inter-domain handover involves some additional procedures such as "authenticating to the new domain." The handover delay remains to be unacceptable even if the registration period is very short.

There are also several schemes proposed to handle the inter-domain handovers. In Kwon et al. 2002, the MN establishes security associations (SAs) with all neighboring domains whenever a session is set up so that neighboring networks can obtain MN's authentication information in advance. When MN roams into one of the neighboring networks, the authentication can be processed locally so that the handover delay can be reduced. A fast handover method is also proposed in Banerjee et al. 2006. In this scheme, the MN reports the new contact address to both C and the proxy in the old domain when it register in the new domain. The proxy in the old domain then set up a temporary session between old proxy and the MN to forward the receiving packets. This session will be terminate until the MN can stably receive the packets from the CN.

3.2.4   Transport Layer Approaches

Transport layer solutions follow the end-to-end principle (Saltzer et al. 1984) in the Internet: anything that can be done in the end system should be done there. Since the transport layer is the lowest end-to-end layer in the Internet protocol stack, it is a natural candidate for vertical handover support. Moreover, in the transport layer solutions, no third party other than the endpoints participates in vertical handover and no modification or addition of network components is required. Snoeren and Balakrishnan 2000 proposed a new set of migrate options for TCP to support mobility. In this protocol, the MN and CN will determine a unique token number to identify the TCP connection when the connection is set up. Such token numbers are calculated according to the address/port numbers of peer nodes. The main drawback of these approaches is that TCP has been globally deployed so that it is practically infeasible to change it. A new IETF-standardized transport layer protocol called Stream Control Transmission Protocol (SCTP) (Stewart 2007b), which can be used in place of both the TCP and UDP, has gained significant attention as a candidate transport protocol for the next generation Internet. The multihoming, multistream and partial reliable data transmission features of SCTP are especially attractive for applications that have stringent performance and high reliability requirements. In Ma et al. 2004, a MN can be configured with multiple addresses in a connection with one address as the primary address. Using SCTP to enable vertical handovers has many advantages, including simpler network architecture, improved throughput and delay performance, and ease of adapting flow/congestion control parameters to the new network during and after vertical handovers (Ma et al. 2004).

3.3 Mobile SCTP for Node Mobility Support

3.3.1 Overview of Mobile SCTP

SCTP has been accepted by the IETF as a general-purpose transport protocol (Stewart 2007b). While it inherits many TCP functions, it also incorporates many attractive new features such as multihoming, multistream and partial reliability. Unlike TCP, which provides reliable in-sequence delivery of a single byte stream, SCTP has a partial ordering mechanism whereby it can provide in-order delivery of multiple message streams between two hosts. This multistream mechanism benefits applications that require reliable delivery of multiple, unrelated data streams, by avoiding head-of-line blocking. The multihoming feature enables an SCTP session to be established over multiple interfaces identified by multiple IP addresses. SCTP normally sends packets to a destination IP address designated as the primary address, but can redirect packets to an alternate secondary IP address if the primary IP address becomes unreachable. Accordingly, the path between two SCTP hosts using their primary addresses is the primary path, and a path between two SCTP hosts using one or more secondary addresses is a secondary path. While only one primary path exists between two SCTP hosts, more than one secondary paths can be available. The set of available connecting paths forms an SCTP association. An SCTP association between two hosts, say, A and B, is defined as:

A set of IP addresses at A + transport port-A + a set of IP addresses at B + transport port-B.

Any of the IP addresses at either host can be used as the corresponding source or destination address in an IP packet sent by one host to the other. Before data can be

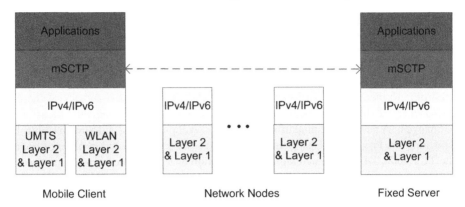

Figure 3.3 Protocol architecture of using mSCTP.

exchanged, the two SCTP hosts must exchange the set of available IP addresses in the association establishment stage. The multihoming mechanism was originally designed for fault-resilient communications between two SCTP endpoints over wired networks. This powerful feature has been exploited to support IP mobility in integrated heterogeneous wireless networks using SCTP.

The SCTP multihoming feature and DAR extension (Stewart 2007a), collectively referred as mobile SCTP (mSCTP) (Koh et al. 2004), can be used to solve the VHO problem in integrated heterogeneous wireless networks at the transport layer by dynamically switching between alternate network interfaces. Since no addition or modification of network components is required, the proposed scheme has a network architecture that is simpler than that of network layer or application layer solutions for VHO.

3.3.2   Protocol Architecture

Figure 3.3 shows the simplified protocol architecture of the proposed scheme. Both MC and FS are assumed to implement mSCTP. In addition, we require both endpoints to implement SCTP message bundling. The MC supports both UMTS and WLAN at the physical and data link layers. There is no additional protocol requirement for other network nodes. To allow access to any FSs over the Internet in general, and recognizing that at the present time these FSs are likely to support TCP rather than mSCTP, the FS in Figure 3.3 can in fact be a proxy server that provide mSCTP associations with MCs over UMTS/WLAN while connecting to other FSs via TCP over the Internet.

3.3.3   Vertical Handover Procedures

Using the multihoming feature of SCTP, an MC can have two IP addresses during vertical handover, one from the UMTS and the other from the WLAN. Similarly, an FS can also be configured for: (1) single-homing, i.e., FS provides only one IP address to support handover; or (2) dual-homing, i.e., FS allows more than one (usually two) IP addresses to support handover. Note that almost all servers in the current Internet are configured with only one IP address. The vertical handover procedures of the single-homing and dual-homing configurations consist of three basic steps: (1) Add IP address; (2) Vertical handover triggering; and 3) Delete IP address. For detailed handover procedures of both single-homing and multihoming configurations, please refer to Ma et al. 2004.

3.4 Performance Improvement of Mobile SCTP for Vertical Handover Support

In this section, we present the SCTP performance degradation problem when used for vertical handover support. Then, we introduce a scheme to improve the performance.

3.4.1 SCTP Behavior Subject to Packet Losses

Mobility between cellular and WLAN is not symmetric. There are two types of cellular/WLAN vertical handovers: unforced handover where a mobile client is covered by both wireless networks and decides to handover form one network to another, and forced handover where a mobile client is leaving the coverage of one network while covered by another network. During a WLAN to cellular forced vertical handover, SCTP performance is affected by two types of packet losses: the dropping of consecutive packets on the WLAN link because of the loss of radio coverage, referred as handover loss, and random packet loss over cellular link due to noise and interference, referred to as error loss. For both handover losses and error losses, the rules for SCTP data transmissions and retransmissions can be summarized as follows: (1) by default, each endpoint should always transmit new data via the primary path; and (2) a sender should try to retransmit a chunk (i.e., an SCTP data unit) to an alternative active destination IP address over the secondary path if transmission over the primary path has failed.

Although the standard SCTP retransmission scheme attempts to improve the chance of success by retransmitting to an alternate peer IP address, it was found that this rule often degrades performance (Caro et al. 2003). The reason is that the retransmission policy specified for SCTP has been developed under the assumption that: (1) a loss indicates that either the current destination IP address is unreachable or the network link is congested; and under such conditions and (2) data transmission condition in the alternate link is better than that in the current primary link. In the WLAN to cellular forced handover scenario, when an MC is moving out of the WLAN coverage and the WLAN link is likely unreachable, all handover data losses on the WLAN link are retransmitted on the cellular link. If one of these packets is lost on the cellular link because of error loss, the packet should be retransmitted via the WLAN link; however, because the WLAN link is likely unreachable, the packet is likely to be kept waiting in the WLAN sending-buffer. While the cellular link is unable to receive any information about this retransmission for a long time, it considers the packet as lost and a time out retransmission is triggered. This results in both WLAN and cellular links going through lengthy time out retransmission processes. It will greatly reduce the data transmission rates and easily cause the SCTP association to shut down.

3.4.2   SMART-FRX Scheme

We extend the improved solution proposed in (Caro et al. 2003) to properly deal with different losses as follows. If the sender detects a loss by D-SACK, it considers the cause to be an error loss, and forces retransmission to the same destination IP address as the original transmission, rather than the alternate destination IP address. If a lost packet causes a time out retransmission at the sender, it considers the loss to be caused by a handover loss, and retransmits the lost packet to the alternative destination IP address. Therefore, when a forced handover occurs on the WLAN link, lost packets due to handover losses are retransmitted over the secondary cellular link; on the other hand, error losses over the cellular link are recovered by retransmitting to the same destination IP address as before.

Our proposed SMART mechanism copies all the buffered data from the primary WLAN link to the secondary cellular link so that the data in sending-buffers of both WLAN and cellular links are completely the same. New data are queued at the tails of both of these two sending-buffers. Because of the copying and multicasting actions, instead of waiting and only retransmitting the time out data from the WLAN link, the cellular link enters into a slow start process since the time of the first time out is detected on the WLAN link. This duplicate queuing action ends when the data sender detects the end of the WLAN possibly unreachable period. After this, (1) if the WLAN link is restored, new data are queued only in the WLAN sending-buffer, the WLAN link remains as the primary connection, and the cellular link becomes idle after all the buffered data have been sent out; or (2) if finally the WLAN link error counter n exceeds PMR, a WLAN to cellular forced VHO is triggered, and subsequently the cellular link becomes the primary connection so that new data are queued only in the cellular sending-buffer, while the WLAN link becomes the secondary connection and goes to the inactive state. In any case, the slow start data transmission process on the cellular link starts at the beginning of the WLAN possibly unreachable period, instead of at the end of this period.

The difference between a handover loss and an error loss is that a handover loss is usually detected by a time out, while an error loss is detected by a sender receiving D-SACKs. Both TCP and SCTP detect losses in two ways, the retransmission timer time outs and D-SACKs (or duplicate ACKs), but with minor difference. In SCTP, a multihomed sender should try to retransmit a missing chunk detected by either a time out or a D-SACK to an alternative active address, while in TCP, the sender does not have this multihoming capability. Except for this difference, SCTP bases its congestion control on the TCP congestion control principles.

In order to avoid error losses on the cellular link being unnecessarily retransmitted on the WLAN link when buffered data are multicast under SMART sub-scheme, as a complement, we propose the FRX sub-scheme, which forces retransmissions of packet losses caused by error loss to the same rather than the alternate destination IP address. According to the current SCTP fast retransmit algorithm, a missed data chunk can be retransmitted quickly on the alternative link before the retransmission timer T3-rtx expires. Because of the shorter loss detection time of D-SACKS compared with that of a time out, the data throughput of recovering from an error loss is generally better than that from a handover loss. If a loss is detected by time out, a handover loss is likely and SMART is activated to multicast the buffered and new data on both primary and alternate IP addresses. If a loss is detected by D-SACKs, an error loss is likely and the FRX sub-scheme is applied whereby the SCTP fast retransmission mechanism is triggered with the retransmission sent to the same destination IP address. With the proposed FRX solution, during a WLAN to cellular forced handover, unnecessary retransmissions on the possibly unreachable WLAN link triggered by D-SACKs on the cellular link, and the resulting long waiting delays can be avoided. An analytical model for SCTP data transmission to study its throughput performance is also presented in Ma et al. 2007.

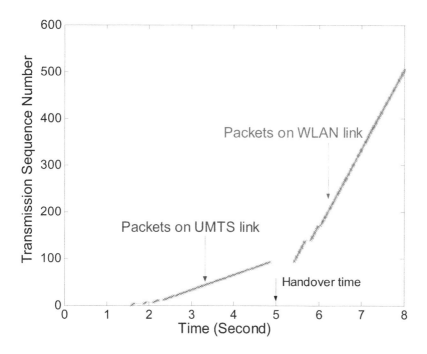

Figure 3.4 Delay performance of SCTP-based vertical handover scheme (single-homing configuration).

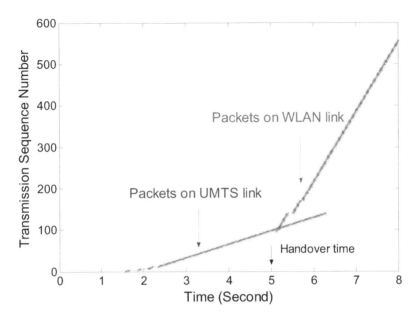

Figure 3.5   Delay performance of SCTP-based vertical handover scheme (dual-homing configuration).

3.5   Simulation Results and Discussions

In this section, we present and discuss the simulation results. Network simulator ns-2 is used in our simulations. The IEEE802.11 WLAN model in ns-2 is used to represent the MAC layer. The bandwidths are set to be 384 kbps for the UMTS link and 2 Mbps for the WLAN link. File transfer protocol (FTP) traffic is simulated. the chunk size is 1468 bytes and the message transmission unit (MTU) size is 1500 bytes. We exam the delay and throughput performance of SCTP in vertical handover support.

   We first assume that there is no packet loss over both WLAN and cellular wireless links. Figs. 3.4 and 3.5 show the delay performance for vertical handover from UMTS to WLAN with single-homing and dual-homing configuration in the FS, respectively. When the FS is in single-homing configuration, the handover delay is the time interval that the FS receives the first packet on the new primary link and the last packet on the old primary link. According to the simulation results, the UMTS toWLAN handover delay is 533 ms in Figure 3.4. When FS is in dual-homing configuration, the handover delay is the time interval that the FS receives the same transmission sequence number on both links. The handover delay is reduced to 234 ms in Figure 3.5. This is because when the FS is in single-homing configuration, the MC sends a 'Set Primary Address' request to trigger a handover, thus increasing the over-

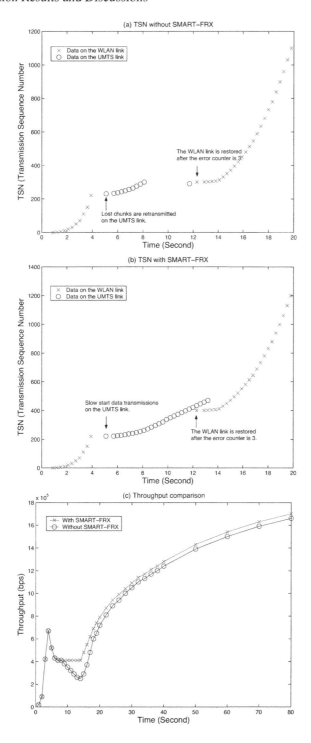

Figure 3.6 Performance of the schemes with and without SMART-FRX.

all delay with a handshake processing time. However, when the FS is in dual-homing configuration, the MC can trigger a handover by directly setting the FS's secondary address; therefore the handover delays in both directions are reduced significantly.

Next, we check the sequence number of each data chunk at the receiver and compare the delay and throughput performance to evaluate the effectiveness of the SMART-FRX scheme. The WLAN link is modeled by a packet-dropping error module that drops consecutive packets to simulate handover losses so that the WLAN link encounters TOs and the maximum error counter n becomes $1, 2, \ldots, 6$, respectively. The cellular link is modeled by a random loss error model to simulate error losses. We show the effectiveness of the proposed SMART-FRX scheme by comparing the throughput performance with and without the proposed scheme. Figure 3.6a shows the received TSNs without SMART-FRX. Figure 3.6b shows the received TSNs with SMART-FRX. Figure 3.6c compares the throughput with and without the proposed scheme. The throughput is defined as the total number of bits received at the receiver divided by the total time elapsed. As shown by Figure 3.6a and Figure 3.6b, at the beginning, the data traffic is on the primary WLAN link. We drop a block of chunks from TSN 214 to 260 on the WLAN link at time 3.9 s, so that the error counter on this link can be set to 1. After about 1 s, i.e., at about 4.95 s, the time out of TSN 214 is detected. Then, on the WLAN link, the cwnd size becomes 1, the T3-rtx increases to two RTO, i.e., 2 s, and the link enters into a failure detection period. Immediately after the error counter becomes non-zero, TSN 214 to 260 are retransmitted on the alternative UMTS link. At the same time, on the WLAN link, with the new T3-rtx and the new cwnd value, TSN 261 is sent to test if the WLAN link has recovered. If we keep on dropping the chunks from TSN 261 to 266, we can make the error counter n to be increased to 2 up to 6. According to our observations on the TSN, with and without SMART-FRX, with the error counter increasing, data transmissions on the WLAN link become disrupted until either of the following happens.

3.6   Conclusions

In next generation heterogeneous wireless networks, the complementary characteristics of different wireless technologies make it attractive to integrated them together. In this chapter, we have presented a SCTP-based mobility support scheme. Protocol architecture and vertical handover procedure are presented. Moreover, we have shown that, due to packet losses in wireless networks, SCTP-based vertical handover scheme suffers from poor performance in forced handover cases. An error recovery scheme has been proposed in this chapter to improve the performance. We have demonstrated the performance of the proposed schemes using simulation results.

Bibliography

Banerjee N, Acharya A and Das SK 2006 Seamless SIP-based mobility for multimedia applications. *Network, IEEE* **20**(2), 6–13.

Caro AL (Jr.), Amer PD, Iyengar JR and Stewart RR 2003 Retransmission policies with transport layer multihoming. *ICON 2003*. pp. 255–260.

Gustaffson E, Jonsson A and Perkins C 2001 Mobile IP Regional Registration.

Gustafsson E and Jonsson A 2003 Always best connected. *Wireless Communications, IEEE* **10**(1), 49–55.

Handley M, Schulzrinne H, Schooler E and Rosenberg J 1999 SIP: Session Initiation Protocol. IETF RFC 2543.

Haverinen H and Malinen J 2001 Mobile IP Regional Paging.

Hsieh R, Zhou ZG and Seneviratne A 2003 S-MIP: a seamless handoff architecture for mobile IP *INFOCOM 2003*, vol. 3, pp. 1774–1784.

Johansson P, Kazantzidis M, Kapoor R and Gerla M 2001 Bluetooth: an enabler for personal area networking. *Network, IEEE* **15**(5), 28–37.

Koh SJ, Lee MJ, Riegel M, Ma L and Tuexen M 2004 Mobile SCTP for Transport Layer Mobility. Internet-Draft draft-sjkoh-sctp-mobility-04.txt, Internet Engineering Task Force.

Kwon TT, Gerla M and Das S 2002 Mobility management for VoIP service: Mobile IP vs. SIP. *Wireless Communications, IEEE* **9**(5), 66–75.

Ma L, Yu F, Leung VCM and Randhawa T 2004 A new method to support UMTS/WLAN vertical handover using SCTP. *Wireless Communications, IEEE* **11**(4), 44–51.

Ma L, Yu FR and Leung VCM 2007 Performance Improvements of Mobile SCTP in Integrated Heterogeneous Wireless Networks. *Wireless Communications, IEEE Transactions* **6**(10), 3567–3577.

Misra A, Das S, Dutta A, McAuley A and Das SK 2002 IDMP-based fast handoffs and paging in IP-based 4G mobile networks. *Communications Magazine, IEEE* **40**(3), 138–145.

Murata Y, Hasegawa M, Murakami H, Harada H and Kato S 2009 The architecture and a business model for the open heterogeneous mobile network. *IEEE Comm. Mag.* **47**(5), 95–101.

Perkins C 2002 IP Mobility Support for IPv4. IETF RFC 3220.

Perkins CE and Johnson DB 2001 Route Optimization in Mobile IP. Internet-Draft draft-ietf-mobileip-optim-11.txt., Internet Engineering Task Force.

Perkins CE and Wang K 1999 Optimized Smooth Handoffs in Mobile IP *ISCC '99: Proceedings of the The Fourth IEEE Symposium on Computers and Communications*, p. 340.

Salkintzis AK 2004 Interworking techniques and architectures for WLAN/3G integration toward 4G mobile data networks. *Wireless Communications, IEEE* **11**(3), 50–61.

Saltzer JH, Reed DP and Clark DD 1984 End-to-end arguments in system design. *ACM Trans. Comput. Syst.* **2**(4), 277–288.

Snoeren AC and Balakrishnan H 2000 An end-to-end approach to host mobility *MobiCom '00: Proceedings of the 6th annual international conference on Mobile computing and networking*, pp. 155–166.

Soliman H, Castelluccia C, El-Malki K and Bellier L 2004 Hierarchical Mobile IPv6 Mobility Management (HMIPv6). Internet-Draft draft-ietf-mipshop-hmipv6-02.txt, Internet Engineering Task Force.

Stewart R 2007a Stream Control Transmission Protocol (SCTP) Dynamic Address Reconfiguration. IETF RFC 5061.

Stewart R 2007b Stream Control Transport Protocol.

Vali D, Paskalis S, Kaloxylos A and Merakos L 2003 An efficient micro-mobility solution for SIP networks *Global Telecommunications Conference, 2003. GLOBECOM '03. IEEE*, vol. 6, pp. 3088–3092.

Wu W, Banerjee N, Basu K and Das SK 2005 SIP-based vertical handoff between WWANs and WLANs. *Wireless Communications, IEEE* **12**(3), 66–72.

Yang L and Giannakis GB 2004 Ultra-wideband communications: an idea whose time has come. *Signal Processing Magazine, IEEE* **21**(6), 26–54.

4

Concurrent Multipath Transfer Using SCTP Multihoming

Janardhan Iyengar

This chapter investigates end-to-end *Concurrent Multipath Transfer* (CMT) using SCTP multihoming for increased application throughput and reliability. CMT is the simultaneous transfer of new data from a source host to a destination host via two or more end-to-end paths. Simultaneous transfer of new data to multiple destination addresses is a natural extension to SCTP multihoming, but there are several design considerations that we explore in this chapter.

4.1   CMT Algorithms

As is the case with TCP, reordering introduced in an SCTP flow degrades throughput. When multiple paths being used for CMT have different delay and/or bandwidth characteristics, significant packet reordering can be introduced in the flow by a CMT sender. Reordering is a natural consequence of CMT, and is impossible to eliminate in an environment where the end-to-end path characteristics are changing or unknown *a priori*, as in the Internet. In this section, we identify and resolve the three negative side-effects of reordering introduced by CMT that must be managed before the performance gains of parallel transfer can be fully achieved: (i) unnecessary fast retransmissions at the sender (Section 4.1.1), (ii) reduced cwnd growth due to fewer cwnd updates at the sender (Section 4.1.2), and (iii) increased ack traffic due to fewer delayed acks (Section 4.1.3). [1]

4.1.1   Preventing Unnecessary Fast Retransmissions - SFR Algorithm

When reordering is observed, a receiver sends gap reports (i.e., gap acks) to the sender which uses the reports to detect loss through a fast retransmission procedure similar to TCP's Fast Retransmit (Allman et al. 1999; R. Stewart 2007). With CMT, unnecessary fast retransmissions can be caused due to reordering, with two negative consequences: (1) since each retransmission is assumed to occur due to a congestion loss, the sender reduces its cwnd for the destination on which the retransmitted data was outstanding, and (2) a cwnd overgrowth problem causes a sender's cwnd to grow aggressively for the destination on which the retransmissions are sent, due to acks received for original transmissions.

Conventional interpretation of a SACK chunk in SCTP (or SACK options in TCP) is that gap reports imply possible loss. The probability that a TSN is lost, as opposed to being reordered, increases with the number of gap reports received for that TSN. Due to sender-induced reordering, a CMT sender needs additional information to infer loss. Gap reports *alone* do not (necessarily) imply loss; but a sender can infer loss using gap reports *and* knowledge of each TSN's destination.

Algorithm Details: The proposed solution to address the side-effect of incorrect cwnd evolution due to unnecessary fast retransmissions is the *Split Fast Retransmit (SFR)* algorithm (Figure 4.1). SFR introduces a *virtual queue* per destination within the sender's retransmission queue. A sender deduces missing reports for a TSN using SACK information in conjunction with state maintained about the transmission destination for each TSN in the retransmission queue. With SFR, a multihomed sender correctly applies the fast retransmission procedure on a *per destination* basis. An advantage of SFR is that only the sender's behavior is affected; the SCTP receiver is unchanged.

SFR introduces two additional variables per destination at a sender:

1. *highest_in_sack_for_dest* - stores the highest TSN acked per destination by the SACK being processed.

2. *saw_newack* - a flag used during the processing of a SACK to infer the causative TSN(s)'s destination(s). Causative TSNs for a SACK are those TSNs

---

[1] These results have been published in Iyengar et al. (2004a, 2006, 2004b).

which caused the SACK to be sent (i.e., TSNs that are acked in this SACK, and are acked for the first time).

On receipt of a SACK containing gap reports [Sender side behavior]:
1) $\forall$ destination addresses d_i, initialize $d_i.saw\_newack$ = FALSE;
2) **for** each TSN t_a being acked that has not been acked
    in any SACK thus far **do**
        let d_a be the destination to which t_a was sent;
        set $d_a.saw\_newack$ = TRUE;
3) $\forall$ destinations d_n, set $d_n.highest\_in\_sack\_for\_dest$ to highest TSN
    being newly acked on d_n;
4) to determine whether missing report count for a TSN t_m
    should be incremented:
        let d_m be the destination to which t_m was sent;
        if $(d_m.saw\_newack$ == TRUE) **and**
        $(d_m.highest\_in\_sack\_for\_dest > t_m)$ **then**
        increment missing report count for t_m;
        else do not increment missing report count for t_m;

Figure 4.1 Split Fast Retransmit (SFR) algorithm – eliminating unnecessary fast retransmissions.

In Figure 4.1, step (2) sets *saw_newack* to TRUE for the destinations to which the newly acked TSNs were sent. Step (3) tracks, on a per destination basis, the highest TSN being acked. Step (4) uses information gathered in steps (2) and (3) to infer missing TSNs. Two conditions in step (4) ensure correct missing reports: (a) TSNs to be marked should be outstanding on the same destination(s) as TSNs which have been newly acked, and (b) at least one TSN, sent later than the missing TSN, but *to the same destination address*, should be newly acked.

4.1.2 Avoiding Reduction in Cwnd Updates - CUC Algorithm

The cwnd evolution algorithm for SCTP (R. Stewart 2007) (and analogously for SACK TCP (Allman et al. 1999; Floyd et al. 2000)) allows growth in cwnd only when a new cumulative ack (cum ack) is received by a sender. When SACKs with unchanged cum acks are generated (say due to reordering) and later arrive at a sender, the sender does not modify its cwnd. This mechanism again reflects the conventional view that a SACK which does not advance the cum ack indicates possibility of loss due to congestion.

Since a CMT receiver naturally observes reordering, many SACKs are sent containing new gap reports but not new cum acks. When these gaps are later acked by a new cum ack, cwnd growth occurs, but only for the data newly acked in the most recent SACK. Data previously acked through gap reports will not contribute to cwnd growth. This behavior prevents sudden increases in the cwnd resulting in bursts of data being sent. Even though data may have reached the receiver "in-order per destination," without changing the current handling of cwnd, the updated cwnd will not reflect this fact.

This inefficiency can be attributed to the current design principle that the cum ack in a SACK, which tracks the latest TSN received in-order at the receiver, applies to an entire association, not per destination. TCP and current SCTP (i.e., SCTP without CMT) use only one destination address at any given time to transmit new data to, and hence, this design principle works fine. Since CMT uses multiple destinations simultaneously, cwnd growth in CMT demands tracking the latest TSN received in-order *per destination*, information not coded directly in a SACK.

While a new separate sequence number space per destination seems like a natural decision at this point, but CMT's design operates under the constraint that it was to modify as little of on-the-wire SCTP as possible, and with modifications to only the sender, if possible.

We propose a cwnd growth algorithm to track the earliest outstanding TSN *per destination* and update the cwnd, even in the absence of new cum acks. The proposed *Cwnd Update for CMT (CUC)* algorithm uses SACKs and knowledge of transmission destination for each TSN to deduce in-order delivery per destination. The crux of the CUC algorithm is to track the earliest outstanding data *per destination*, and use SACKs which ack this data to update the corresponding cwnd. In understanding our proposed solution, we remind the reader that gap reports alone do not (necessarily) imply congestion loss; SACK information is treated only as a concise description of the TSNs received by the receiver.

Algorithm Details: Figure 4.2 shows the proposed CUC algorithm. A *pseudo-cumack* tracks the earliest outstanding TSN per destination at the sender. An advance in a pseudo-cumack triggers a cwnd update for the corresponding destination, even when the actual cum ack is not advanced. The pseudo-cumack is used for cwnd updates; only the actual cum ack can dequeue data in the sender's retransmission queue since a receiver can renege on data that is not cumulatively acked. An advantage of CUC is that only the sender's behavior is affected; the SCTP receiver is unchanged.

The CUC algorithm introduces three variables per destination at a sender:

1. *pseudo_cumack* - maintains earliest outstanding TSN.

2. *new_pseudo_cumack* - flag to indicate if a new pseudo-cumack has been received.

3. *find_pseudo_cumack* - flag to trigger a search for a new pseudo-cumack. This flag is set after a new pseudo-cumack has been received.

In Figure 4.2, step (2) initiates a search for a new *pseudo_cumack* by setting *find_pseudo_cumack* to TRUE for the destinations on which TSNs newly acked

---

At beginning of an association [Sender side behavior]:
    $\forall$ destinations d, reset
        $d.find\_pseudo\_cumack$ = TRUE;
On receipt of a SACK [Sender side behavior]:
    1) $\forall$ destinations d, reset $d.new\_pseudo\_cumack$ = FALSE;
    2)   **if** the SACK carries a new cum ack **then**
           for each TSN t_c being cum acked for the first time, that was not
           acked through prior gap reports **do**
               (i) let d_c be the destination to which t_c was sent;
               (ii) set $d_c.find\_pseudo\_cumack$ = TRUE;
               (iii) set $d_c.new\_pseudo\_cumack$ = TRUE;
    3)   **if** gap reports are present in the SACK **then**
           for each TSN t_p being processed from the retransmission queue **do**
               (i) let d_p be the destination to which t_p was sent;
               (ii) **if** $(d_p.find\_pseudo\_cumack == \text{TRUE})$ **and**
                  t_p was not acked in the past **then**
                      $d_p.pseudo\_cumack = t_p$;
                      $d_p.find\_pseudo\_cumack$ = FALSE;
               (iii) **if** t_p is acked via gap reports for first time **and**
                  $(d_p.pseudo\_cumack == t_p)$ **then**
                      $d_p.new\_pseudo\_cumack$ = TRUE;
                      $d_p.find\_pseudo\_cumack$ = TRUE;
    4)   **for** each destination d **do**
           if $(d.new\_pseudo\_cumack == \text{TRUE})$ **then** update cwnd as
           per R. Stewart (2007);

Figure 4.2 Cwnd Update for CMT (CUC) algorithm – handling side-effect of reduced cwnd growth due to fewer cwnd updates.

were outstanding. A cwnd update is also triggered by setting *new_pseudo_cumack* to TRUE for those destinations. Step (3) then processes the outstanding TSNs at a sender, and tracks on a per destination basis, the TSN expected to be the next *pseudo_cumack*. Step (4) finally updates the cwnd for a destination if a new *pseudo_cumack* was seen for that destination.

4.1.3   Curbing Increase in Ack Traffic - DAC Algorithm

Sending an ack after receiving every two data PDUs (i.e., delayed acks) in SCTP (and TCP) reduces ack traffic in the Internet, thereby saving processing and storage at routers on the ack path. SCTP specifies that a receiver should use the delayed ack algorithm as given in RFC 2581 (Allman et al. 1999), where acks are delayed only as long as the receiver receives data in order. Reordered PDUs should be acked immediately. With CMT's frequent reordering, this rule causes an SCTP receiver to frequently *not* delay acks. Hence a negative side-effect of reordering with CMT is increased ack traffic.

To prevent this increase, we propose that a CMT receiver ignore the rule mentioned above. That is, a CMT receiver does not immediately ack an out-of-order PDU, but delays the ack. Thus, a CMT receiver always delays acks, irrespective of whether or not data are received in order.[2] Though this modification eliminates the increase in ack traffic, RFC 2581's rule has another purpose which gets hampered.

According to RFC 2581, a receiver should immediately ack data received above a gap in the sequence space to accelerate loss recovery with the fast retransmit algorithm (Allman et al. 1999). In SCTP, four acks with missing reports for a TSN indicate that a receiver received at least four data PDUs sent after the missing TSN. Receipt of four missing reports for a TSN triggers the sender's fast retransmit algorithm. In other words, the sender has a *reordering threshold* (or *dupack threshold*) of four PDUs. Since a CMT receiver cannot distinguish between loss and reordering introduced by a CMT sender, the modification suggested above by itself would cause the receiver to delay acks even in the face of loss. When a loss does occur with our modification to a receiver, fast retransmit would be triggered by a CMT sender only after the receiver receives eight(!) data PDUs sent after a lost TSN – an overly conservative behavior.

The effective increase in reordering threshold at a sender can be countered by reducing the actual number of acks required to trigger a fast retransmit at the sender, i.e., by increasing the number of missing reports registered per ack. In other words, if a sender can increment the number of missing reports more accurately per ack received, fewer acks will be required to trigger a fast retransmit. A receiver can provide more information in each ack to assist the sender in accurately inferring the number of missing reports per ack for a lost TSN. We thus propose that in each ack, a receiver report the number of data PDUs received since the previous ack was sent. A sender then infers the number of missing reports per TSN based on the TSNs being acked in a SACK, number of PDUs reported by the receiver, and knowledge of transmission destination for each TSN.

Algorithm Details: The proposed *Delayed Ack for CMT (DAC)* algorithm (Figure 4.3) specifies a receiver's behavior on receipt of data, and also a sender's behavior when the missing report count for a TSN needs to be incremented. Since SCTP (and TCP) acks are cumulative, loss of an ack will result in loss of the data PDU count reported by the receiver, but the TSNs will be acked by the following ack. Receipt of this following ack can cause ambiguity in inferring missing report count per destination. Our algorithm conservatively assumes a single missing report count per destination in such ambiguous cases. The DAC algorithm requires modifications to both the sender and the receiver.

No new variables are introduced in DAC, as we build on the SFR algorithm. An

---

[2]We do not modify a receiver's behavior when an ack being delayed can be piggybacked on reverse path data, or when the delayed ack timer expires.

On receipt of a data PDU [Receiver side behavior]:
  1) delay sending ack as given in R. Stewart (2007), with the additional rule that the ack should be delayed even if reordering is observed.
  2) in ack, report number of data PDUs received since sending of previous ack.

When incrementing missing report count through SFR:Step (4)
(Figure 4.1) [Sender side behavior]:
  4) to determine whether missing report count for a TSN t_m should be incremented:
     let d_m be the destination to which t_m was sent;
     if ($d_m.saw\_newack$ = TRUE) **and**
     ($d_m.highest\_in\_sack\_for\_dest > t_m$) **then**
         (i) **if** ($\forall$ destinations d_o where $d_o \neq d_m$, $d_o.saw\_newack$ == FALSE)
         then
               (all newly acked TSNs were sent to the same destination as t_m)
               (a) **if** ($\exists$ newly acked TSNs t_a, t_b such that $t_a < t_m < t_b$) **then**
                   (conservatively) increment missing report count for t_m by 1;
               (b) **else if** ($\forall$ newly acked TSNs t_a, $t_a > t_m$) **then**
                   increment missing report count for t_m by number of PDUs reported by receiver;
         (ii) **else**
               (Mixed SACK - newly acked TSNs were sent to multiple destinations)
               (conservatively) increment missing report count for t_m by 1;

Figure 4.3 Delayed Ack for CMT (DAC) Algorithm – Handling side-effect of increased ack traffic.

additional number is reported in the SACKs for which we propose using the first bit of the flags field in the SACK chunk header - 0 indicates a count of one PDU (default SCTP behavior), and 1 indicates two PDUs.

In Figure 4.3, at the receiver side, steps (1) and (2) are self-explanatory. The sender side algorithm modifies step (4) of SFR, which determines whether the missing report count should be incremented for a TSN. DAC dictates *how many* to increment by. Step (4-i) checks if only one destination was newly acked, and allows incrementing missing reports by more than one for TSNs outstanding to that destination. Further, all newly acked TSNs should have been sent later than the missing TSN. If there are newly acked TSNs that were sent before the missing TSN, step (4-i-a) conservatively increments missing reports by only one. If more than one destinations are newly acked, step (4-ii) conservatively increments by only one.

These three algorithms—SFR, CUC, and DAC—constitute the core of CMT's (mostly) sender-based algorithms for distributing traffic over multiple paths. We now explore retransmission policies for CMT in the following section.

4.2 Retransmission Policies

Multiple paths present an SCTP sender with several options where to send a retransmission. But the choice is not well-informed since SCTP restricts sending new data, which can act as probes for information (such as available bandwidth, loss rate and RTT), to only one primary destination. Consequently, an SCTP sender has minimal information about paths to a receiver other than the path to the primary destination.

On the other hand, a CMT sender maintains accurate information about all paths, since new data are being sent to all destinations concurrently. This information allows a CMT sender to better decide where to retransmit. That is, a CMT sender can choose the retransmission destination for each loss recovery independently using the most recent information for its decision.

We present five retransmission policies for CMT (Iyengar et al. 2004a). In four policies, a retransmission may be sent to a destination other than the one used for the original transmission. Previous research on SCTP retransmission policies shows that sending retransmissions to an alternate destination degrades performance primarily because of the lack of sufficient traffic on alternate paths **?**. With CMT, data are concurrently sent on all paths, thus the results in **?** are not applicable. The five retransmission policies for CMT are:

- **RTX-SAME** - Once a new data chunk is scheduled and sent to a destination, all retransmissions of that chunk are sent to the same destination (until the destination is deemed *inactive* due to failure **?**).

- **RTX-ASAP** - A retransmission of a data chunk is sent to any destination for which the sender has cwnd space available at the time of retransmission. If multiple destinations have available cwnd space, one is chosen randomly.

- **RTX-CWND** - A retransmission is sent to the destination for which the sender has the largest cwnd. A tie is broken by random selection.

- **RTX-SSTHRESH** - A retransmission is sent to the destination for which the sender has the largest ssthresh. A tie is broken by random selection.

- **RTX-LOSSRATE** - A retransmission is sent to the destination with the lowest loss rate path. If multiple destinations have the same loss rate, one is selected randomly.

Of the policies, RTX-SAME is simplest. RTX-ASAP is a "hot-potato" policy - retransmit as soon as possible without regard to loss rate. RTX-CWND and RTX-SSTHRESH practically track, and attempt to move retransmissions onto the path with the estimated lowest loss rate. Since ssthresh is a slower moving variable than cwnd, the values of ssthresh may better reflect the conditions of the respective paths. RTX-LOSSRATE uses information about loss rate provided by an "oracle" - information that RTX-CWND and RTX-SSTHRESH estimate. This policy represents a hypothetically ideal case; hypothetical since in practice, a sender typically does not know *a priori* path loss rates; ideal since the path with the lowest loss rate has highest chance of having a packet delivered. We hypothesized that retransmission policies that take loss rate into account would outperform ones that do not.

Most of these policies allow a retransmission to be sent to a different destination than the original transmission. We now discuss two protocol modifications that are required to allow redirecting retransmissions to a different destination than the original.

4.2.1   CUCv2: Modified CUC Algorithm

The CUC algorithm (Figure 4.2) enables correct cwnd updates in the face of increased reordering due to CMT. To recap Section 4.1.2, the CUC algorithm recognizes a set of TSNs outstanding per destination, and the per-destination *pseudo_cumack* traces the left edge of this list of TSNs, per destination. CUC assumes that retransmissions are sent to the same destination as the original transmission. The per-destination *pseudo_cumack* therefore moves whenever the corresponding left edge is acked; the TSN on the left edge being acked may or may not have been retransmitted.

If the assumption about the retransmission destination is violated, and a retransmission is made to a different destination from the original, CUC cannot faithfully track the left edge on either destination. We modify CUC to permit the different retransmission policies. The modified algorithm, named CUCv2 is shown in Figure 4.4.

CUCv2 recognizes that a distinction can be made about the TSNs outstanding on a destination - those that have been retransmitted, and those that have not. CUCv2 maintains two left edges for these two sets of TSNs - *rtx_pseudo_cumack* and *pseudo_cumack*. Whenever either of the left edges moves, a cwnd update is triggered.

At beginning of an association [Sender side behavior]:
   ∀ destinations d, reset
       $d.find\_pseudo\_cumack = d.find\_rtx\_pseudo\_cumack$ = TRUE;
On receipt of a SACK [Sender side behavior]:
   1)   ∀ destinations d, reset
           $d.new\_pseudo\_cumack = d.new\_rtx\_pseudo\_cumack$ = FALSE;
   2)   **if** the ack carries a new cum ack **then**
           for each TSN t_c being cum acked for the first time, that was
           not acked through prior gap reports **do**
               (i) let d_c be the destination to which t_c was sent;
               (ii) set $d_c.find\_pseudo\_cumack$,
                   $d_c.find\_rtx\_pseudo\_cumack$,
                   $d_c.new\_pseudo\_cumack$,
                   $d_c.new\_rtx\_pseudo\_cumack$ to TRUE;
   3)   **if** gap reports are present in the ack **then**
           for each TSN t_p being processed from the retransmission queue **do**
               (i) let d_p be the destination to which t_p was sent;
               (ii) **if** ($d_p.find\_pseudo\_cumack$ == TRUE)
                   and t_p was not acked in the past
                   and t_p was not retransmitted **then**
                       $d_p.pseudo\_cumack = t_p$;
                       $d_p.find\_pseudo\_cumack$ = FALSE;
               (iii) **if** t_p is acked via gap reports for first time
                   and ($d_p.pseudo\_cumack == t_p$) **then**
                       $d_p.new\_pseudo\_cumack$ = TRUE;
                       $d_p.find\_pseudo\_cumack$ = TRUE;
               (iv) **if** ($d_p.find\_rtx\_pseudo\_cumack$ == TRUE)
                   and t_p was not acked in the past **and** t_p was
                   retransmitted **then**
                       $d_p.rtx\_pseudo\_cumack = t_p$;
                       $d_p.find\_rtx\_pseudo\_cumack$ = FALSE;
               (v) **if** t_p is acked via gap reports for first time
                   and ($d_p.rtx\_pseudo\_cumack == t_p$) **then**
                       $d_p.new\_rtx\_pseudo\_cumack$ = TRUE;
                       $d_p.find\_rtx\_pseudo\_cumack$ = TRUE;
   4)   **for** each destination d **do**
           if ($d.new\_pseudo\_cumack$ == TRUE)
           or ($d.new\_rtx\_pseudo\_cumack$ == TRUE) **then**
               update cwnd as per **??**;

Figure 4.4 CUCv2 algorithm - modified Cwnd update for CMT (CUC) algorithm.

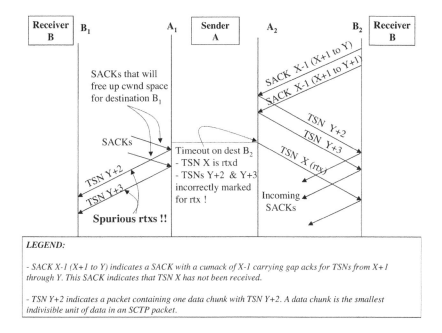

Figure 4.5 Example of spurious retransmissions after timeout in CMT.

4.2.2 Spurious Timeout Retransmissions

When a timeout occurs, an SCTP sender is expected to bundle and send as many of
the earliest TSNs outstanding on the destination for which the timeout occurred as
can fit in an MSS (Maximum Segment Size) PDU. Per RFC 2960, more TSNs that
are outstanding on that destination "should be marked for retransmission and sent as
soon as cwnd allows (normally when a SACK arrives)". This rule is intuitive. While
sending, retransmissions are generally given priority over new transmissions. As in
TCP, the cwnd is also collapsed to 1 MSS for the destination on which a timeout
occurs.

A timeout retransmission can occur in SCTP (as in TCP) for several reasons.
One reason is loss of the fast retransmission of a TSN. Consider Figure 4.5. When
a timeout occurs due to loss of a fast retransmission, some TSNs that were just sent
to the destination on which the timeout occurred are likely awaiting acks (in Figure 4.5, TSNs Y+2 and Y+3). These TSNs get incorrectly marked for retransmission
on timeout. With the different CMT retransmission policies, these retransmissions
may be sent to a different destination than the original transmission. In Figure 4.5,
spurious retransmissions of TSNs Y+2 and Y+3 are sent to destination B_1, on receipt
of acks freeing up cwnd space for destination B_1. Spurious retransmissions are exac-

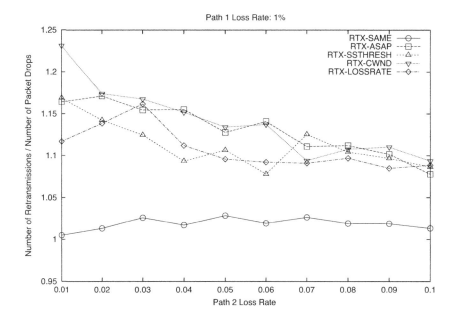

Figure 4.6 Spurious retransmissions in CMT without RTT heuristic.

erbated in CMT, as shown through this illustration, due to the possibility of sending data (including retransmissions) to multiple destinations concurrently.

We simulated the occurrence of such spurious retransmissions with the different retransmission policies in CMT. Figure 4.6 shows the ratio of retransmissions relative to the number of actual packet drops at the router. Ideally, the two numbers should be equal; all curves should be straight lines at $y = 1$. Figure 4.6 shows that spurious retransmissions occur commonly in CMT with the different retransmission policies.

We propose a heuristic to avoid these spurious retransmissions. Our heuristic assumes that a timeout cannot be triggered on a TSN until the TSN has been outstanding for at least one RTT. Thus, if a timeout is triggered, TSNs which were sent within one RTT are not marked for retransmission. We use an average measure of the RTT for this purpose - the smoothed RTT, which is maintained at a sender. This heuristic requires the sender to maintain a timestamp for each TSN indicating the time at which the TSN was last transmitted (or retransmitted). Figure 4.7 shows how the application of this heuristic dramatically reduces spurious retransmissions.

Performances of the policies differ significantly; especially when limited receive buffer sizes are considered in the next section.

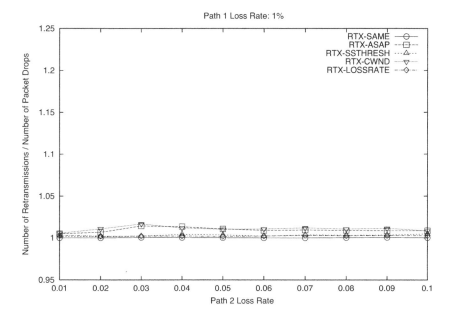

Figure 4.7 Spurious retransmissions in CMT with RTT heuristic.

4.3 Socket Buffer Blocking

A transport layer receiver maintains *receive socket buffer* space for incoming data for two reasons: (i) to handle out-of-order data, and (ii) to receive data at a rate higher than that of the receiving application's consumption. In SCTP (and TCP), a receiver advertises currently available *rbuf* space through window advertisements (normally accompanying acks) to a data sender. Similarly, a sender maintains *send socket buffer* space to hold on to data that has not yet been *cumulatively* acknowledged by the receiver, so that this data may be retransmitted if a loss is detected by the sender.

Irrespective of the layer at which multipath transfer is performed, similar shared buffers would exist at a receiver (likely at the transport or application layer). These buffers degrade overall throughput. We will consider both these buffers, and methods to alleviate the degradation in this section.

4.3.1 Degradation due to Receive Buffer Constraints

Figure 4.8a shows the time taken for a CMT sender to transfer an 8MB file when the *rbuf* is set to 64KB, using the five retransmission policies. RTX-SAME is the simplest to implement, but performs worst. The performance difference between RTX-SAME and other policies increases as the loss rate on Path 2 increases. RTX-ASAP performs better than RTX-SAME, but still worse than RTX-LOSSRATE, RTX-

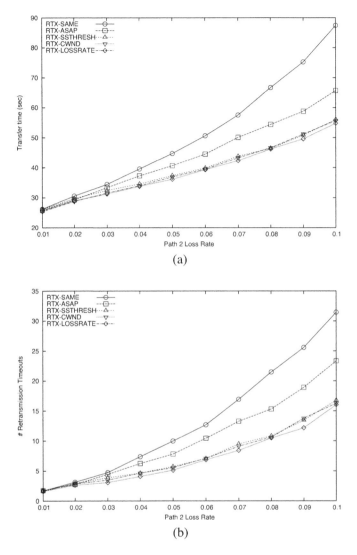

Figure 4.8 *rbuf* = 64KB, and Path 1 loss rate = 1%: (a) CMT time to transfer 8MB file, (b) Retransmission timeouts for CMT with different policies.

SSTHRESH and RTX-CWND. The three loss rate-based policies perform equally.

Figure 4.8b shows the number of retransmission timeouts experienced when using the different policies. This figure shows that performance improvement in using RTX-LOSSRATE, RTX-CWND, and RTX-SSTHRESH is due to the reduced number of timeouts. A lost transmission may be recovered via a fast retransmission, but a lost fast retransmission can be recovered only through a timeout. RTX-SAME does

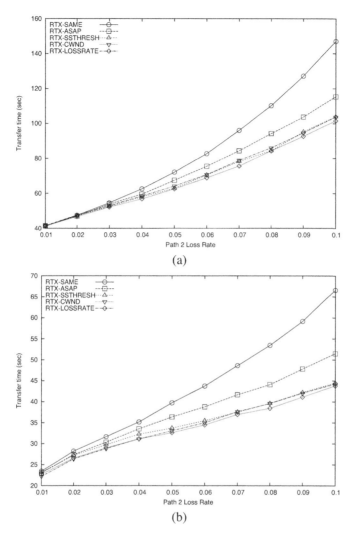

Figure 4.9 Path 1 loss rate = 1%, CMT time to transfer 8MB file using: (a) *rbuf*=32KB, (b) *rbuf*=128KB.

not consider loss rate in choosing a retransmission destination and consequently experiences the largest number of timeouts due to increased loss of retransmissions.

Figures 4.9a and b show performance of the retransmission policies with *rbuf* sizes of 32KB and 128KB respectively. Together with Figure 4.8a, we can see that the smaller the *rbuf*, the more important the choice of retransmission policy.

Figures 4.8 and 4.9 show that *rbuf* size has a strong impact on CMT performance. When CMT is used over paths with different loss rates, a constrained rbuf that is shared within an association causes performance degradation since data dropped on a higher loss rate path causes data received on lower loss rate paths to be blocked from delivery to the receiving application (assuming ordered delivery.) We call this phenomenon *rbuf blocking*. Degradation increases with a reduction in *rbuf* size, and/or an increase in the number of timeouts (Iyengar et al. 2006).

Using loss-rate-based policies alleviates *rbuf* blocking since the number of timeouts is reduced. From Figure 4.8a and Figure 4.9, as *rbuf* size decreases, *rbuf* blocking increases, and loss-rate-based policies perform increasingly better than the other policies. (See Iyengar et al. 2006 for an extensive discussion of *rbuf* blocking.)

Figures 4.8 and 4.9 suggest that any retransmission policy that takes loss rate into account will likely improve load distribution for both new transmissions and retransmissions. Retransmissions will be redirected to a lower loss rate path, avoiding inactive timeout recovery periods, and allowing new transmissions to be sent on the higher loss rate path, thus maintaining a flow of data on both paths. Policies that take loss rate into account avoid repeated retransmissions and timeouts - thus also improving the timeliness of data.

A general conclusion we draw from our evaluations is that *loss-rate-based policies are equally the best performing policies*. We thus recommend RTX-CWND and RTX-SSTHRESH, the practical loss-rate-based policies, for CMT. We note that careful choice of retransmission policy can significantly reduce performance degradation due to *rbuf* blocking.

4.3.2   Degradation due to Send Buffer Constraints

TCP and SCTP both employ two kinds of data acknowledgment mechanisms: (i) cumulative acks indicate data that has been received in-sequence, and (ii) selective acknowledgments (SACKs) indicate data that has been received out-of-order. While cumulatively acked data are the receiver's responsibility, SACKed data are not. SACK information is considered advisory only (to allow receiver reneging), and a sender is expected to retain all data that is not cumulatively acknowledged.

Thus, a constrained send buffer (*sbuf*) can cause degradation in throughput since a sender with a blocked *sbuf* will be unable to send any data until cumulative acks open up *sbuf* space at the sender. This degradation is the same as the *rbuf* degradation that we discussed earlier in this section, and can be alleviated by careful choice of retransmission policy.

Interestingly, while performance degradation with *rbuf* blocking is reduced when multistreaming is used—head-of-line blocking is reduced—*sbuf* blocking dominates when multistreaming is used. This is because *sbuf* blocking is independent of the number of streams used in the association and depends purely on the association-wide TSNs. In addition to the retransmission policies, we have proposed a new SACK chunk, called Non-Renegable SACK (NR-SACK), that significantly reduces this degradation due to *sbuf* blocking.

In summary, both send and receive socket buffers are important for performance, but their impact is higher with CMT since they effectively couple the performance of the individual paths used for CMT.

Bibliography

Allman M, Paxson V and Stevens W 1999 TCP Congestion Control. RFC2581.

Floyd S, Mahdavi J, Mathis M and Podolsky M 2000 An Extension to the Selective Acknowledgement (SACK) Option for TCP. RFC2883.

Iyengar J, Amer P and Stewart R 2004a Retransmission Policies For Concurrent Multipath Transfer Using SCTP Multihoming *ICON 2004*, Singapore.

Iyengar J, Amer P and Stewart R 2006 Concurrent Multipath Transfer Using SCTP Multihoming Over Independent End-to-End Paths. *IEEE/ACM Transactions on Networking*.

Iyengar J, Shah K, Amer P and Stewart R 2004b Concurrent Multipath Transfer Using SCTP Multihoming *SPECTS 2004*, San Jose, California.

Stewart R 2007 Stream Control Transmission Protocol. RFC4960.

5

Low Delay Communication and Multimedia Applications

Eduardo Parente Ribeiro

This chapter focuses on low delay communication methods that SCTP can provide for real-time multimedia applications. The basic strategy to select a path with lower delay for transmission is described. Most works use SCTP internal variable SRTT (smoothed round-trip time) as estimation of the current delay on each path. It is calculated based on acknowledgments received (data and heartbeat). Other parameters such a delay threshold, guard time, losses are also employed to prevent unnecessary route changes. Quality improvements under specific scenarios are evaluated and several examples are shown. Asymmetric communication aspects are described and commented. Discussions and future directions finishes the chapter.

5.1   Introduction

One important measure that can be optimized in a multihomed scenario is end-to-end packet delay. This is particularly relevant for real-time multimedia applications. Multimedia data are information conveyed in form of video, audio, image or a combination of them. Multimedia applications can be divided in three classes regarding the users' need for interaction. First class includes files that are transmitted to be used later. There is no need to immediate interaction. Second class includes application like video and audio on demand. There is a limited need for interaction in terms of pause, rewind, skip, etc. A fairly large amount latency are accepted for those actions. They can take a few seconds to take place. The third class of multimedia applications are related to interaction that must occur in real-time, like voice conversation or video-conferencing. There is need for a sub-second synchronization between the two parts involved in the conversation. The exact amount of accepted delay varies according to users expectation. ITU recommends value of 100 ms for maximum network delay for realtime high-interaction applications (VoIP, Videoconferencing) (ITU-T 2006).

There is an increasing demand for all types of multimedia communication. From the communication network point of view the class of interactive multimedia applications presents a greater challenge because of the stringent time limits requirements and large delays variations that data networks tend to exhibit. Buffers can be employed to accommodate delay variations but they are more suited for on-demand video and audio applications since extra delay is introduced.

Although other mechanisms of SCTP presented on previous chapters are still applicable to all multimedia transmission in a general form they are not specifically suited for realtime multimedia communication where the tight time constraints plays an important role on perceived quality by the end user.

Concurrent Multi-Transfer (CMT) is a good approach for bulk transfers but it does not necessarily provide the low delay and low jitter requirements for realtime applications. Voice transmission does not require high throughput (rate is between 5 to 80 kbps) but packet must arrive in timely fashion. Video conferencing would require some more bandwidth depending on the quality employed but low packet delay and jitter are the prominent desired feature. It is possible to imagine some scenarios where CMT would not be a good solution for delay sensitive applications. When one of the available paths have intrinsically higher delay than the other (e.g., satellite link or a path traversing many intermediate routers) it is not beneficial to use it for transmission concurrently with other faster path. Some packet would experience long delays unnecessarily and jitter would be increased significantly as well.

The failover mechanism of standard SCTP is certainly an important method to provide resilience for an established data communication session. But the time it takes for this standard approach to effectively switch transmission to alternate path is very high. It depends on the protocol parameter Path Maximum Retransmissions (PMR) which indicates the number of retransmission that can be carried out on a path. Thus PMR+1 retransmissions will denote a path failure and cause further transmission to use an alternate idle path. Protocol recommended PMR value of 5 results in a time to failure of more than one minute. Lower PMR values can be used (0 to 4) and failure detection time can be reduced significantly, but the possibility of spurious failovers increases accordingly. Caro Jr. et al. (2004) have investigated performance of various PMR values (0 to 5) and concluded that PMR=0 performs well for simulated bulk transfers. Different policies for retransmission have been proposed. Three of them were evaluated in another work by Caro Jr. et al. (2006): All Retransmissions to Same path (AllRtxSame), All Retransmissions to Alternate path (AllRtxAlt) and Fast Retransmissions to Same path, Timeouts to Alternate path (FrSameRtoAlt), which attempts to balance the tradeoffs between the other two. The results obtained with network simulator for file transfer scenarios indicated policies perform differently depending on the network conditions. FrSameRtoAlt presented a good tradeoff balance.

Realtime multimedia applications need a very fast handover time. Ideally, the handover on a ongoing session should be such that the receiver application would receive all packets in due time during this process. An analytical estimation of the failover time in SCTP multihoming scenarios is provided by Budzisz et al. (2007).

5.1.1 Audio and Video Coding

The initial stage for digital transmission of video and audio is the coding and decoding process. This is accomplished by the coder and decoder (Codec) block. The analog information of video and audio needs to be converted to a digital representation and coded using the least number of bits to save transmission bandwidth. The process of coding can be simple and fast usually yielding a higher bit rate or complex and slow allowing smaller rates. Compression algorithms usually rely on psycho-visual or psycho-acoustic model of human perception to balance the number of bits where the received information will be more or less accurately perceived. Some speech coders employ other approach by performing a parametric coding of the voice source based on some predetermined mathematical model of voice formation. Small bit rates can be achieved (around 2 kbps) at the expense of loss of natural voice sounding. Speech Codecs can be divided into 3 classes: waveform, source and hybrid coders. The latter attempt to fill the gap between waveform and source coder. They allow better speech quality with a smaller increase in bit rate (to around 10 kbps) compared to source coders. A large number of Codecs are in use nowadays. Some of them are proprietary. ITU has many standards published for audio and video codecs. Recommendation H.263 ITU-T (2005) defines protocols to provide audio/visual communication over packet-based networks. It relies on a series of other recommendations that specifies codecs and call control procedures. A brief description of popular Codecs is given below.

1. Voice

   G.711 ITU-T (1988) is largely used in conventional telephony and also still used in some voice over IP (VoIP) applications due to its simplicity. It is a waveform coder where signal is sampled at 8 kHz and each sample is logarithmic compressed to be efficiently represented by 8 bits. This pulse-coded modulation (PCM) stream has a constant bit rate of 64 kbps. Other examples of waveform coders are G.722 and G.726 that uses adaptive differential PCM (ADPCM) to reduce the bit rate to 32 or even 16 kbps with a small decrease in speech quality. Examples of hybrid codecs are specified in G.729, G.729a and G.722.2. They employ Algebraic Code Excited Linear Prediction (ACELP) to yield good speech quality at low bit rate (6 to 23 kbps). These codecs are typically used in mobile or VoIP communication. G.729a is compatible with G.729 but requires less computation. G.722.2 implements Adaptive Multi-Rate Wideband (AMR-WB) algorithm and provides wider voice bandwidth 50–7000 Hz when compared to other codecs designed for conventional telephony (300–3400 Hz). Those speech codecs usually operate at constant bit rate but some codecs can use silence suppression to save bits or even choose to code the voice signal with varying number of bits yielding a variable bit rate (VBR). This means that lower bit rate can be achieved for the same speech quality or conversely more quality can be obtained for the same mean bit rate.

2. Video/Audio

Video codecs which usually operates with variable bit rate. ITU provides a family of video coding standards. H.261 ITU-T (1993) is an old standard and designed to support two frame sizes: CIF (common interchange format – 352x288) and QCIF (quarter CIF – 176 x 144). It could operate at video bit rates from 40 kbps to 2 Mbps. H.263 ITU-T (2005) was an evolution from H.261 and was largely deployed in videoconferencing systems. The most recent Standard is H.264 ITU-T (2009) called Advanced video coding for generic audiovisual services. It is intended to cover a broad range of applications such as videoconferencing, digital storage media, television broadcasting and Internet streaming. It was designed in close collaboration with ISO/IEC Moving Picture Experts Group (MPEG) where it is named MPEG-4 Part 10 (ISO/IEC 2008). The standards are jointly maintained, they have identical technical content. MPEG-4 was first released in 1998 and absorbed many features of previous standards MPEG-2 and MPEG-1. They specify a collection of methods for compressing both video and audio. The compression of the audio part of the movie was initially standardized by MPEG-1 which provided three layers with increased compression level (MP1, MP2 and MP3 – the latter gained wide acceptance for audio storage and transfer becoming the popular coding method for portable players). MP3 was further enhance in MPEG-2 specification which also introduced new coding method called Advanced Audio Coding (AAC) sought to be the successor of MP3. MPEG-4 uses AAC.

Table 5.1 MOS: quality and impairment scales

	Quality		Impairment
5	Excellent	5	Imperceptible
4	Good	4	Perceptible, but not annoying
3	Fair	3	Slightly annoying
2	Poor	2	Annoying
1	Bad	1	Very annoying

Table 5.2 Recommendations for subjective assessment of quality

Category	Recommendation
Voice	ITU-T P.800 Methods for objective and subjective assessment of quality
Video	ITU-R BT.500 Methodology for the subjective assessment of the quality of television pictures ITU-T P.910 Subjective video quality assessment methods for multimedia applications
Audio	ITU-R BS.1284 General methods for the subjective assessment of sound quality (audio alone) ITU-R BS.1286 Methods for the subjective assessment of audio systems with accompanying picture ITU-T P.911 Subjective audiovisual quality assessment methods for multimedia applications

5.1.2 Assessing the Quality: MOS, E-Model (ITU) and Others

Media degradation that occurs during transmission due to packet loss can be determined by comparing the received signal with its original version. There are several ways to express this error. Mean squared error (MSE), Root Mean Squared Error (RMSE) and Peak Signal to Noise Rate (PSNR) are some examples. This objective measure of quality does not always reflect the true degree of perceived quality.

Perceived quality of the received media is a subjective matter and it is considered statistically in terms of the average of individual opinions. Mean opinion Score (MOS) provides a numerical indication of the perceived quality. Table 5.1 shows the scale that goes from 1 (bad) to 5 (excellent) regarding the quality, or 1 (very annoying) to 5 (imperceptible) regarding the impairment (ITU-T 1996). The assessment of quality via MOS for several medias is standardized by ITU recommendations as described in Table 5.2.

MOS assessment is a time-consuming and expensive procedure. There are however objective calculations that consider some forms of psycho-visual-acoustic elements of human perception. Those are called perceptual evaluation methods. Many

Table 5.3 Speech transmission user satisfaction

E-Model Rating	quality	category	MOS
$90 \leq R < 100$	Best	Very satisfied	$4.3 \leq MOS < 100$
$80 \leq R < 90$	High	Satisfied	$4.0 \leq MOS < 4.3$
$70 \leq R < 80$	Medium	Some users dissatisfied	$3.6 \leq MOS < 4.0$
$60 \leq R < 70$	Low	Many users dissatisfied	$3.1 \leq MOS < 2.6$
$50 \leq R < 60$	Poor	Nearly all users dissatisfied	$1.0 \leq MOS < 2.6$

methods have been proposed and this theme is still a matter of discussion. Some methods currently in use are:

PESQ – Perceptual evaluation of speech quality (ITU-T 1998).

PEVQ – Perceptual evaluation of Video Quality (ITU-T 2008c).

PEAQ – Perceptual evaluation of Audio Quality (ITU-R 2001).

A simple but not so accurate way to assess speech quality is obtained with E-model (ITU-T 2008b). This is a computation model to help transmission planners to build systems ensuring users will be satisfied with end-to-end transmission performance. It has been adapted from conventional telephony systems to VoIP transmissions. E-model relates the impairments due to several factors such as noise, codecs, network delay and jitter to provide one figure of merit named R rating according to

$$R = R_0 - I_s - I_d - I_e + A \tag{5.1}$$

where R_0 represents the transmission impairment based on the signal-to-noise ratio, I_s is the effect of impairments to the voice signal, I_d is the effect of impairments due to delay, I_e is the degradation of quality caused by low bit rate codecs and A is a compensation factor based on user expectation. A simple calculation tool and tutorial is provided by ITU-T (2008a). R rating can be converted to the MOS scale using

$$MOS = \begin{cases} 1 & R < 0 \\ 1 + 0.035R + 7R(R - 60)(100 - R)10^{-6} & 0 \leq R \leq 100 \\ 4.5 & R > 100 \end{cases} \tag{5.2}$$

Table 5.3 describes range of values for R, its meaning, and equivalent MOS.

Although E-Model does not takes into account the type of delay distribution experienced by the packets nor differentiate burst losses from uniform losses which might provide more accurate estimation it is simple, straightforward and fast to calculate. It is a useful tool for comparing VoIP quality in different simulated scenarios (Santos et al. 2007).

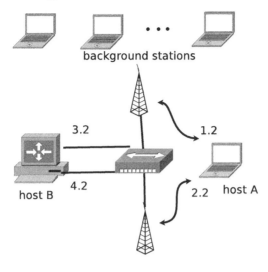

Figure 5.1 Equipment setup (Kelly et al. 2004).

5.2 Delay-Centric Strategy: Switch Transmission Path for Low Latency

After the publication of SCTP standard (RFC2600) in 2000 many works began considering this new protocol as a good alternative to improve the transport of multimedia traffic (Caro Jr. et al. 2001; Kashihara et al. 2003). A simple idea to use path delay to perform a handover with SCTP was proposed by researchers of Performance Engineering Lab at University College Dublin (Kelly et al. 2004). They considered a WLAN scenario where path latency may degrade due to several factors including traffic congestion. An estimation of the active path delay is obtained by using SCTP internal variable SRTT which is a low pass filtered version of the instantaneous RTT. When an ACK chunk is received, SCTP updates SRTT value for the active path according to

$$\text{SRTT} = (1 - \alpha)\text{SRTT} + \alpha\text{RTT} \tag{5.3}$$

where $\alpha = 0.125$. The inactive paths are probed less frequently by HB chunks. The interval between two probes is given by: $H_i = RTO_i + HB.interval(1 + \delta)$ where RTO_i is the latest RTT time-out value for destination i, and δ is a random value between -0.5 and 0.5. The standard parameter HB.interval is 30 s. RTO_i usually has a small value and δ serves to introduce some variability to the probe times. The RTT measurement on the secondary paths is considered merely a guide to the expected RTT if data traffic were to be carried on that path. In order to demonstrate the proposed scheme a WLAN scenario showed on Figure 5.1 was setup (Kelly et al. 2004).

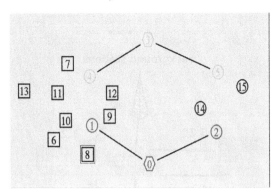

Figure 5.2 Wireless topology of the experiment (Noonan et al. 2004).

Two multihomed SCTP hosts running Linux operating system were used as the endpoints of the association. They modified standard SCTP linux implementation available in (Stewart and Xie 2001) for delay-centric handover to one of the hosts only. No changes needed to be made to the second host providing a form of backward compatibility to standard SCTP. Host A can communicate to host B through two separate wireless networks with different IP address ranges. Once the association is established traffic flows on the primary path (0) to host B.

After some seconds several wireless stations start transmitting UDP packets to increase background traffic. This caused congestion in the cell and led to an increase in path delay. When they stop transmitting congestion ceases and delay on primary path returns to baseline level. This example illustrated the possibility of simple delay-centric strategy to perform a vertical handover on a congested network to a less congested one and back again to the first network when congestion ended.

Other work by the same group (Noonan et al. 2004) presented controlled simulations to further demonstrate the technique they also called Delay Sensitive SCTP (DS-SCTP). A network Simulator (NS2) was used to investigate some wireless scenarios. Figure 5.2 shows a representation of the employed topology. Wireless nodes 0 and 3 are multihomed. They can communicate through interfaces 1-4 over network A or interfaces 2-5 over network B. Scenario 1 represents a badly performing wireless LAN where Network A has 8 interfering sources, while network B has only 2. A voice application transmitting from node 0 to 3 over network A is started at 5 seconds, and at about 9 seconds, it is shown that DS-SCTP hands over to the least loaded network in order to improve performance. Another scenario simulates a condition where both networks are lightly loaded with only 2 interfering sources. The load is divided quite evenly (5:4 ratio) between the two networks.

The experiment is run several times for 120 seconds while a hysteresis parameter is varied. A handover is programmed to occur only when average delay on the idle path plus the hysteresis is greater than the average delay on the current path.

Figure 5.3 shows the network used for transmission each time for a 20 millisec-

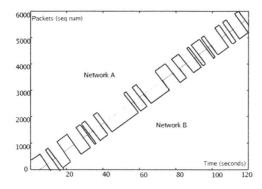

Figure 5.3 Network selected for transmission (Noonan et al. 2004).

Table 5.4 Increasing hysteresis with lightly loaded networks

Hysteresis (ms)	# Handovers	Ratio A:B
0	69	1 :1.25
4	56	1 :0.88
10	43	1 :1.42
20	28	1 :0.96
40	14	1:114

Source: Noonan et al. 2004.

ond hysteresis. The superimposed square wave indicates the network used for transmission.

As summarized in Table 5.4, hysteresis reduces the number of handovers and does not seem to affect the ability to change paths when needed.

A last scenario investigates a changing condition. Network A has 3 sources of traffic throughout the experiment. The load on network B starts with one source and every 15 s another source is added. Network B is used initially but once it becomes congested network A is mostly selected.

The authors remarks that this scheme could be used to allow a user to select between a number of available networks, depending on which was able to offer the best level of service. Multimedia applications are often more sensitive to delay variations and this could be significant when networks use different types of technology. They plan to study the influence of delay and jitter in future work.

The path selection method (PSM) proposed by Kashihara et al. (2003) uses a similar approach. Their selection algorithm considers path latency based on SRTT but also the bottleneck bandwidth (BBW), which is obtained by packet pair measurement. The mobile host sends two HB packets consecutively, and BBW is calculated using the difference between the arrival times of the two packets (ΔT), as given by

$$BBW = \frac{\text{HB packet size}}{\Delta T}$$

They show the result of a simulation on NS2 where a mobile host roam within an overlap area between networks. A VoIP transmission at 64 kbps using SCTP switches to backup path when there is a decrease in the main path bandwidth. One problem with this approach is that bandwidth estimation based on packet pair is notably inaccurate (Prasad et al. 2003). This may lead to erroneous path change or prevent a path change when it is necessary. The idea of on-the-fly bandwidth estimation is very interesting and more investigation on dynamic scenarios may prove it robust and advantageous.

A more accurate estimation of path bandwidth is obtained by monitoring the path throughput. TCP Westwood (Mascolo et al. 2001) employ this method to adjust its transmission window. The same principle was applied to SCTP for a smart load balancing among the available paths (Fiore and Casetti 2005). At the beginning of the association round-robin is used to uniformly distribute the chunks among the paths, until a bandwidth estimation is obtained for every path. Then the chunks to be transmitted are distributed among the paths concurrently but in proportion to path bandwidth. The authors show some simulations on NS2 where this strategy performs better or equal than a simple concurrent multipath transmission. The idea is to avoid sending data over a channel as soon as some room in its congestion window is freed to prevent reordering delays at the receiver. They remark that packets sent over slower channels can arrive at the receiver much later than those sent over faster channels. This yields poor quality in sound and image display as well as large duplicate SACKs transmissions.

A similar approach that estimates path bandwidth but select only one path for transmission is proposed by Fracchia et al. (2007). Bandwidth on primary path is estimated by the ratio between the amount of transmitted data and RTT. On secondary paths the HB packet is replaced by a train of 6 packets (2 small, 2 large, 2 small) and their dispersion is used to estimate bottleneck bandwidth. Because only one path is in use at a time eventual head-of-line blocking that may occurs in CMT is avoided. The proposed technique also uses throughput on the primary path to differentiate losses due interferences in the wireless transmission or due to packet drop on router congested queue. One can speculate approaches based on bandwidth will not work well for cases where path with higher bandwidth has intrinsic high delay (e.g., satellite links) compared to low latency paths with much smaller but sufficient bandwidth. The proposed method is though a good solution for bulk transfers and may perform well for realtime multimedia communication in many scenarios.

The absolute delay experienced by packets may have different consequences to the perceived quality depending on the type of Codec in use. Fitzpatrick et al. (2006) propose another metric to decide the handover: An estimation of user perceived MOS which is continuously calculated using E-Model. The method has the advantage of taking into account not only packet delays but also packet losses. They perform some simulations with an 802.11b WLAN scenario to show that an online estimation of

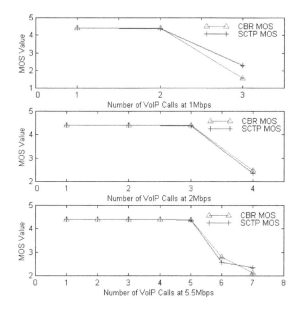

Figure 5.4 CBR and SCTP MOS for G.711 at different data rates (Fitzpatrick et al. 2006).

MOS is similar to its offline calculation. The original heartbeat mechanism was modified to send multiple packets to imitate a VoIP traffic of a G.711 codec. A train of 25 heartbeat packets is sent to each endpoint in the association every T seconds. The chunks were set to have size of 80 bytes transmitted every 10 ms. If packets were excessively delayed they were considered lost. The loss rate was calculated considering a delay threshold of 300 to 350 ms (twice the maximum one way delay with the addition of encoding and decoding delays).

The estimated MOS and the MOS calculated for a CBR traffic were compared. The results are displayed in Figure 5.4 for different numbers of VoIP calls that represents the load of the network. It was verified that a good agreement of both values occurs for all three wi-fi rates. MOS estimation for G.729 codec was also adequate.

This proposed scheme has a good potential of assessing more accurately the perceived user quality. There is the overhead of transmitting a train of packets which needs to be weighted. Nonetheless, the HBs need to be sent only to the alternate paths because the primary path may have MOS estimated from its own VoIP packets acknowledgments. This idea were further developed in later works where the authors called it ECHO: Quality of Service-based Endpoint Centric Handover scheme for VoIP (Fitzpatrick et al. 2009, 2008).

A relevant question that needs to be answer for wide deployment of such delay-centric selection method is about the overall system stability when a large number of users would be using the same strategy to select minimum-delay path. Would the

Table 5.5 Stabilization of path switching regime for different number of SCTP competing agents and path capacity / bandwidth ratio

Agents	C/B ratio							
	1.000	1.050	1.070	1.080	1.090	1.095	1.100	2.000
6	≈	≈	≈	≈	≈	≈	∼	=
12	≈	≈	≈	≈	≈	≈	∼	=
24	≈	≈	≈	≈	∼	≈	=	=
48	≈	≈	≈	≈	≈	=	∼	=

overall system utilization converge to a stable, low-delay situation for all users? The end systems may detect and select an alternate path with low delay but if they all change to this path it would be quickly congested. Another path may be selected by all of them and the end systems would keep oscillating between congested paths causing overall communication to exhibit high latency.

An initial study has shown that this may happen under some circumstances (Gavriloff 2009). A simulated scenario on NS2 considered SCTP sources transmitting VoIP traffic of G.711 Codec. Two hosts are dual-homed to two ISP represented by two routers. There are two distinct paths between these two routers. Each SCTP agent estimates path latency through its internal SRTT variable. SRTT is updated upon receipt of acknowledgment in the active path using the standard parameter $\alpha = 0.125$. On the alternate path heartbeats (HB) are sent every second and HB-ACK updates SRTT of this path accordingly. All agents are initiated on first path and will switch path if they detect that second path has smaller latency. They are randomly started over the first 5 seconds of simulation to avoid synchronous operation. Link capacities on each path (C) needs to be equal or greater than half of total aggregate traffic bandwidth (B). The C/B ratio varied from 1 to 2 by adjusting the capacity for a given traffic bandwidth. Simulations considered 6, 12, 24 and 48 SCTP agents.

Three types of behavior were observed depending on simulation parameters. Figure 5.5 shows the number of agents on the first path during the 250 s of simulation. The number of agents on the second path is simply the complement to 6. First plot (a) shows the case where the agents keep switching paths during the whole simulation and the system never stabilizes. This occurs when there is no or just a small slack on the capacity. Second plot (b) shows the case where agents switch paths for a long time but they stabilize evenly after a while. Third plot (c) shows the case where agents switch paths and quickly stabilize to an even distribution among the two paths. This is the case when the link capacities were not so tight compared to traffic bandwidth.

Table 5.5 shows whether path switching stabilizes for several C/B ratios and different numbers of SCTP agents. These results obtained by visual inspection suggests that a slack of around 10% is necessary to prevent unstable behavior for 6 SCTP agents. For greater number of competing SCTP sources stabilization occurs with smaller C/B ratios.

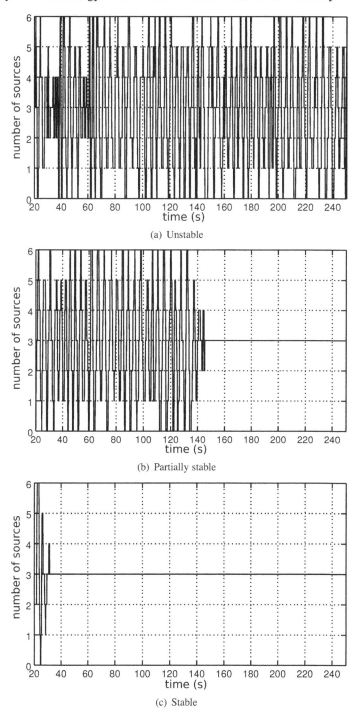

Figure 5.5 Path selection oscillations.

Table 5.6 Stabilization of path switching regime with guard time mechanism

Agents	C/B ratio							
	1.000	1.050	1.070	1.080	1.090	1.095	1.100	2.000
6	≈	≈	≈	≈	≈	≈	≃	=
12	≈	≈	=	=	=	=	=	=
24	≈	≈	≈	≈	=	=	=	=
48	≈	=	=	=	=	=	=	=

In order to help overall system stability and prevent that the agents keep changing paths a delay-hysteresis parameter was considered. An agent switches path only if secondary path SRTT is smaller than current path SRTT less the hysteresis value. This approach did not displayed significant differences.

Another mechanism was proposed and investigated. Guard time is a period an agent must wait to switch path. If anytime during this interval the current path presents lower latency the switch operation is canceled. To further prevent syncronous operation among the competing agents a new random value is considered as the guard time for each agent when it detect lower delay on the alternate path. A uniform time distribution between 0 and 3 s was used. This mechanism showed some small improvement toward stability as can be verified in Table 5.6.

Another measure of quality that was investigated was the estimated perceived user quality mapped to MOS calculated by E-model. An unstable behavior induces higher latency for the VoIP traffic and degrades voice quality. Figure 5.6 shows the average MOS for 10 simulations as a function of C/B ratio.

Low quality indicated by small MOS values occurs for tight capacity values. This is an indication that path oscillations were responsible for quality degradation. When the number of SCTP sources is increased better MOS is obtained. This is because less oscillations were observed but also because higher path capacity imply smaller transmission delays for each packet. Those results suggest that stability should not be a big problem for most scenarios but it may happen in limited circumstances that involves high utilization of path capacity by small number of competing SCTP streams.

The same investigation (Gavriloff 2009) has also showed some results about the perceived user quality of one SCTP VoIP flow over stochastic background traffic. A simple scenario with two dual homed hosts was considered as displayed in Figure 5.7.

A Markovian process where packets inter-arrival time is give by exponential distribution is considered. In this mathematically well know model (M/D/1 in Kendall notation) packet delay in the system is given by

$$\bar{d} = \bar{s} + \frac{\rho \bar{s}}{2(1 - \rho)}$$

where ρ is the link utilization (ration between traffic average bandwidth and link capacity). Because the link transport both this background and SCTP (CBR) traffic,

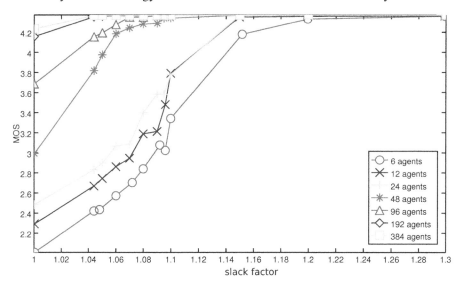

Figure 5.6 Calculated MOS for different number of competing SCTP transmissions versus slack factor

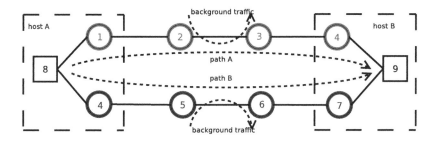

Figure 5.7 Topology used on simulation.

actual utilization is described by: $\rho = \rho_B + \rho_S$, where ρ_B is the utilization due to background traffic ρ_S is the utilization due to VoIP transmission simulated G.711 codec. Figure 5.8 shows the average delay as a function of link utilization. Aggregated traffic composed by CBR transmission and background traffic displays smaller delays when compared to pure M/D/1 system. Link capacity of 500 kbps was considered. Confidence intervals of 95% were obtained after the results from 30 simulations.

The same averaged delay was specified for both paths. One experiment with standard SCTP transmitting on only one path was run for comparison. Figure 5.9 shows the computed MOS for both cases: single-path and multi-path. Even though both paths on the long run have the same mean delay they are uncorrelated and for short

Low Delay Communication and Multimedia Applications

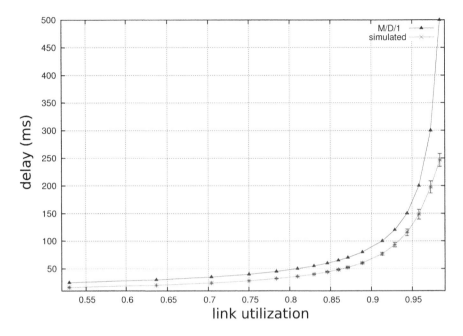

Figure 5.8 Averaged delay versus system utilization.

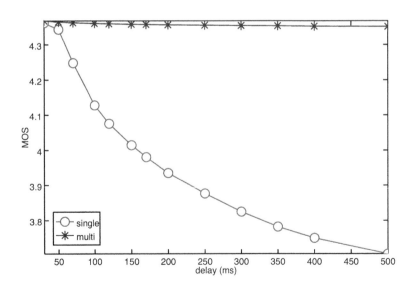

Figure 5.9 Calculated MOS versus delay for multipath and single path.

periods of time one of the paths has smaller delay than the other. Path selection algorithm based on SRTT was able to select the lowest delays instants of each path keeping SCTP overall latency low. The resulting MOS was always high while single path transmission quality degrades for increased mean delay. It can be remarked that this is an specific result for Markovian traffic (M/D/1) which may not be representative of real background traffic pattern. Real traffic tends to be more variable and bursty and exhibits fractal characteristics. Long tail distributions are frequently used to model this behavior. Nevertheless it is expected that delay-centric selection algorithm would perform well or even better for traffics with higher variability.

5.3 Asymmetric Round Trip Path Approach for One-Way Delay Optimization

The strategy to select the lowest delay path can be further refined if one-way delay is considered in the place of round-trip delay. Historically, most delays estimations are based on the simplifying assumption that forward and return delays are the same and equal to half the round-trip time (RTT) of a route. This is rarely true because of the asymmetry present on most traffic profiles of internet applications, e.g., web surfing, file download and video streaming. Access technology like ADSL (Asymmetric Digital Subscriber Line) try to accommodate such imbalance by providing higher bit rate in the downstream direction.

Different delays on forward and return paths have been a concern for TCP congestion control (Barakat et al. 2000). TCP delay estimation for each side is based on RTT measurements. One cannot distinguish if an increase in delay is due to the forward or reverse path. This may lead to under-utilization of the available bandwidth. This problem has been addressed by TCP extensions (Fu and Liew 2003) that change the way congestion window is calculated.

Multihoming introduces a whole new perspective in the sense that not only one-way delay over the available paths can be compared but an action can be taken to select the most appropriate path for transmission in each direction, independently. The first effect of this strategy is to increase the number of paths combinations that can be utilized and the chances of lower delay communication. A collateral effect is that by using the least congested paths in each direction this asymmetric communication may be helping to 'fill-in' the bandwidth gaps induced by other applications and contributing to balance the overall delay on both directions. This could be considered beneficial for TCP and other applications that rely on symmetric RTT that might be running over the same paths.

The approach to use asymmetric paths for low delay communication has been proposed recently (Ribeiro and Leung 2005, 2006). It was shown how one-way delays can be compared in order to select the lowest-delay path for transmission in each direction. An example with forward and reverse path names is displayed in Figure 5.10 where hosts H1 and H2 are multihomed to 3 and 2 ISPs, respectively. Circles 1, 2 and 3 represent the network interfaces of H1, while 4 and 5 represent the network interfaces of H2. It is assumed that packets can be transmitted freely between

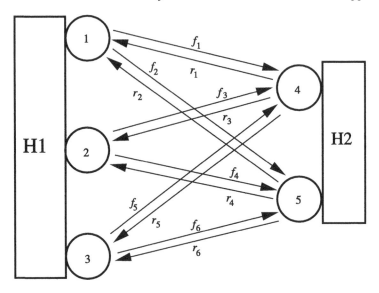

Figure 5.10 Multihoming with asymmetric paths. Forward and reverse paths are identified.

any pair of interfaces (nodes) of the two hosts. Each forward path (relative to H1) is designated f_i and the corresponding reverse path is designated r_i, where $i = 1,\ldots,P$ ($P = 6$ in Figure 5.10) spans the one-way paths existing in each direction. Transmission delay over the corresponding one-way path is d_{f_i} (or d_{r_i}).

Latency on the round-trip paths could be estimated by sending special probing packets that are returned by the receiver via a specific return path (e.g, via heartbeat chunks (HB), possibly extended to specify the return paths for the replies).

To determine relative delays for the forward paths, not all P^2 RTT combinations are required, but a set of P probes along different forward paths returned via the same reverse path are sufficient. The difference in RTTs of paths $f_i r_k$ and $f_j r_k$ yields the delay difference between forward paths f_i and f_j:

$$
\begin{aligned}
\text{RTT}(f_i r_k) - \text{RTT}(f_j r_k) &= d_{f_i} + d_{r_k} - (d_{f_j} + d_{r_k}) \\
&= d_{f_i} - d_{f_j}, \qquad i \neq j; \quad i,j = 1,\ldots,6
\end{aligned}
\tag{5.4}
$$

Forward path with lowest latency is determined by comparing their relative latencies. The same strategy can be applied by the other host (H2) to determine its forward path with lowest latency.

Table 5.7 displays an example of hypothetical one-way delays in milliseconds for each forward and reverse path. Each cell is the sum of forward and reverse delays representing the path round trip time. Forward path delays vary from 10 (f_6) to 800 ms (f_1) while reverse path delays vary from 50 (r_1) to 1400 ms (r_6). Shaded cells in the main diagonal give RTTs of the symmetric two-way paths. A simple delay-centric path selection method that measures only symmetric RTTs will choose the

Table 5.7 One-way delays and all RTT combinations in ms

			reverse path					
			r_1	r_2	r_3	r_4	r_5	r_6
		one-way delay	50	100	400	800	1200	1400
forward path	f_1	800	850	900	1200	1600	2000	2200
	f_2	300	350	400	700	1100	1500	1700
	f_3	100	150	200	500	900	1300	1500
	f_4	40	90	140	440	840	1240	1440
	f_5	20	70	120	420	820	1220	1420
	f_6	10	60	110	410	810	1210	1410

lowest among these shaded values and pick path $f_2 r_2$ which has a RTT of 400 ms. However, packets would experience a 300 ms delay over the forward path, which may exceed the delay requirement of some real-time multimedia application. The proposed asymmetric algorithm will make host H1 to select forward path f_6 which has the lowest delay value (10 ms) among the forward paths and host H2 to select r_1 which has the lowest delay (50 ms) among the reverse paths. Resulting RTT will be 60 ms.

Another example of the advantage of considering asymmetric paths is illustrated by figures 5.11–5.13 (Ribeiro and Leung 2005). This scenario considers two hosts multihomed through two access networks. Forward paths are designated by lower-case letters a,b,c & d while reverse paths are designated by the corresponding up-percase letters A, B, C & D. Time evolution of forward and reverse path delays are displayed on Figures 5.11a and 5.11b. They are simply linearly changing values with added Gaussian random noise over a 60 s time interval sampled every second.

Each side tries to use its lowest delay forward path. It is supposed that host H1 samples its forward path latency every 2 s, while host H2 do it with 4 s interval starting at 2 s. Figure 5.12a shows what would be the selected forward paths (circles) by H1 and the smallest delay path at every second (dots). Transmission starts on path d, then at t=9 s changes to path c, and finally at t=32 s changes to path b. Notice that the smallest delay path (dot) is not selected all the times because H1 does not sample the path every second. On the same figure it can be verified that the experienced delay closely follows the minimum available delay. Figure 5.12b shows the same type of plot for the reverse path, or the forward path with respect to H2. Figure 5.13 compares delays for the all symmetric round-trip paths with asymmetric path. Lower delays can be obtained during most part of the transmission.

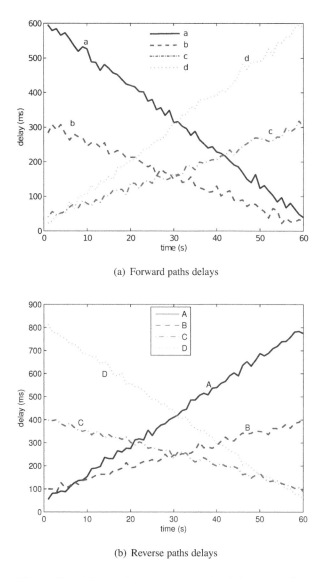

(a) Forward paths delays

(b) Reverse paths delays

Figure 5.11 Forward and reverse paths delays over time.

This example illustrated how SCTP can benefit from asymmetric paths communication when delays on forward and reverse directions are not the same. Another consequence is that the number of round trip paths to choose from are higher. This increases the chances to find a noncongested path.

The number of asymmetric round-trip paths that can be used for communication deserves some discussion. Consider the example in Figure 5.10 where host H1 is

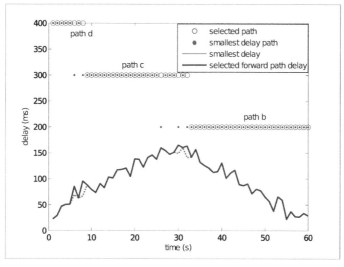

(a) Selected forward paths

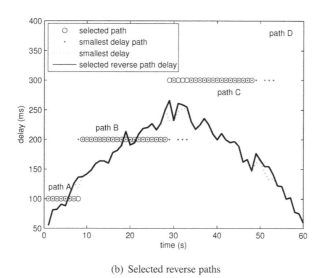

(b) Selected reverse paths

Figure 5.12 Paths delays and selected paths.

multihomed with $M = 3$ interfaces and host H2 is multihomed with $N = 2$ interfaces. There are $P = M \times N = 3 \times 2 = 6$ possible one-way paths in each direction.

Although all P paths are virtually possible, only a subset of that can actually be used in standard SCTP/IP implementation. When SCTP in host H1 wants to send a packet to host H2 it hands over the packet to layer-3 for delivery. Because routing is based on destination address exclusively, IP layer can only send the packet to one of

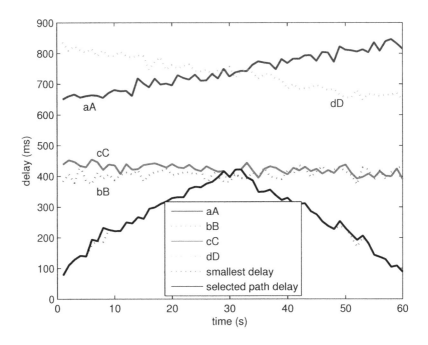

Figure 5.13 Selected round-trip paths and delays.

the two possible addresses (4 or 5). The output interface and hence source address to be used is determined by its routing table which is fixed and previously set. Let us suppose that the output source addresses were, respectively, 1 for destination address 4, and 2 for destination address 5. On the other side, when host H2 wants to send a packet to host H1 it would have 3 possible destination addresses to send the packet. If the paths to 1 and 2 are simultaneously down for some reason, communications between H1 and H2 will break because the routing table in host H1 does not know how to send a packet to host H2 through its third interface 3 as remarked by Stewart and Xie (2001). This example suggests that the number of different one-way paths that actually could be used is $\min(M,N)$. The number of symmetric round-trip paths is the same.

SCTP alone should be easily modifiable to work with asymmetric paths without any change to layer-3. The number of possible symmetric and asymmetric round-trip path combinations would be $\min(M,N)^2$. In this example there are $\min(3,2)^2 = 2^2 = 4$ round trip paths that could be used considering a standard destination routing IP layer. They could be $f_1 r_1, f_6 r_6, f_1 r_6$ and $f_6 r_1$ for a particular routing table setup.

An optimization on SCTP/IP stack would allow layer-4 to tell layer-3 which outgoing interface to use. Then all possible one-way paths $(M \times N)$ in each direction could be used. There is a total of $(M \times N)^2$ possible round-trip path combinations or

Table 5.8 Relative gain of round-trip options compared to standard SCTP

	standard stack	selectable output
Symmetric round-trip paths	1	N
Symmetric and Asymmetric round-trip paths	N	M^2N

$(3 \times 2)^2 = 36$ (!) in this example.

The basic idea of exploring asymmetric round-trip communication is to have more options to choose from in order to find a less congested one-way path. This could be accomplished with standard IP implementation, which gives $(M \times N)$ total round-trip paths, or with selectable output stack, which gives $(M \times N)^2$ options. Table 5.8 summarizes the relative gain of round-trip path options when asymmetry is also considered for both cases. For simplicity $M \geq N$ is assumed.

In a mixed scenario where only one end system has the stack optimization it is easy to verify that the number of possible round-trip path combinations is M^2N when only the stack in host H1 is optimized or MN^2 when only the stack in host H2 is optimized.

The RTT probes could be obtained in different ways. (1) A control packet can be transmitted and the elapsed time to receive the corresponding acknowledgment gives the RTT. (2) During normal operation of SCTP, SRTT are constantly being updated on the primary communication path and less frequently on alternate paths when heartbeats are ACKed. (3) The Heartbeat (HB) mechanism could be slightly changed to suit the needs of this method. The modification required is to respond to all the arriving HBs by returning ACKs over a common return path (e.g., primary path) and not to the incoming address. It is important to note that the main function of HBs was not to probe for path delays originally but to test if a destination address is active (its default interval is 30 s). A delay probe nonetheless, can estimate whether a given address is inactive if it times out before receiving a reply. Both functions are very similar in nature. They could share a common implementation or a separate control chunk may be used for delay probes.

5.3.1 Simulations

Version 2.29 of network simulator (NS2) (*Network Simulator 2* n.d.) was used to test the asymmetric path selection method. Standard SCTP implementation (Caro and Iyengar 2005) has been modified to support asymmetric communications. The heartbeat mechanism has been modified to allow for probing all the possible forward paths to the destination. Instead of replying to the source IP address that originated the HB, all HB-ACKs go from current interface to the primary destination IP address of the correspondent node. This way all the packets take the same return path and allow RTT comparisons at the sender.

The topology used on simulations is illustrated on Figure 5.14. Both sender and receiver are multihomed with two interfaces. The routers are interconnected in a way that all cross-combinations of source-destination addresses between the hosts are possible. The symmetric round-trip paths are: f_1r_1 (nodes 6-8), f_2r_2 (nodes 7-9),

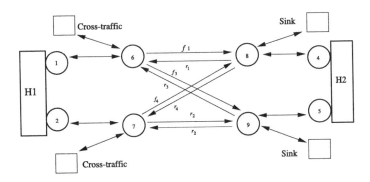

Figure 5.14 Network topology in simulations - SCTP and cross-traffic.

f_3r_3 (nodes 6-9) and f_4r_4 (nodes 7-8). Other asymmetric round-trip may be selected as well. All links have a capacity of 2 Mbps and a fixed delay of 1 ms. Drop-tail queues are used and the length of each queue is set to 750 kB thus allowing enough queue space to prevent packets from being dropped.

The simulated scenario involved only CBR cross-traffic at different rates to establish variable queue occupancy and hence to provide dissimilar delays along the available paths. Figure 5.15 shows the delays experienced by cross-traffic UDP packets when traversing the intermediate routers. This is a baseline scenario for comparison where only cross-traffic is present and SCTP transmission is not active.

Paths f_1, f_2 and f_3 starts with their queue not empty and background traffic at this path has a deficient CBR rate that causes their queue occupancies to decrease as time progresses. Forward path f_4 starts with lowest queue occupancy corresponding to a delay of 20ms at t=0, but background CBR traffic progressively fills up the queue to 200ms at t=50s. Note that at around t=16ms, f_3 becomes the lowest delay forward path as can be seen in Figure 5.15a. Reverse path r_1 begins with its queue almost empty (10 ms delay at t=0) and has a growing delay up to 100 ms delay at t=50 s. Path r_4 has an initial delay of 300ms but its queue occupancy decreased with time ending up with 10ms delay at the end of the simulation. At around t=38 s path r_4 becomes the lowest delay reverse path, as can be seen on Figure 5.15b. SCTP packet size was 100B and packet interval was 0.4 s giving a data rate of 2 kbps. SCTP association started with nodes 1 and 4 set as primary addresses thus using path f_1. HBs to probe for path delays were sent every 2s and instantaneous RTT comparisons were used to select the forward path.

A comparison with standard SCTP is illustrated by Figure 5.16: (1) standard SCTP would transmit only to the primary destination and would not change its destination since there were no path failures. SCTP packets would experience a delay from 200 ms increasing to 230 ms represented by solid squares. (2) Symmetric delay-centric SCTP would have the same behavior because in this particular case

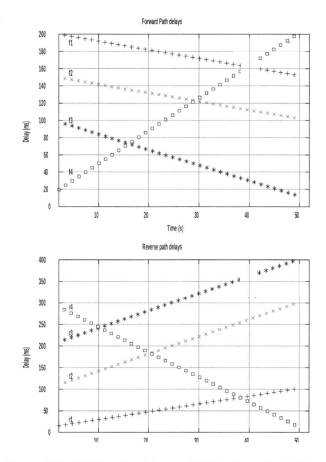

Figure 5.15 UDP packet delays - CBR traffic only. (a) Forward paths, (b) Reverse paths.

primary round-trip path $f_1 r_1$ has the lowest RTT. (3) Asymmetric-path SCTP experiences lower transmission delays of 25 to 60 ms for most part of the communications as shown by the circles. After the first HB, at t=2.2, path f_4 is detected as having the smallest one-way delay and hence selected for transmission. Around t=12 s path f_3 becomes the lowest delay path. As soon as this is detected by the next HB probe around t=14 s it is selected for transmission.

The proposed minimum-delay asymmetric SCTP is aimed at real-time applications which requires latency minimization. Although no considerations have been given to bandwidth availability, this method may show some benefits even when there is some demand for bandwidth. The variable delays that packets experience during transmission are due to queue occupancy on intermediate routers either by packets from cross-traffic or by previous packets from the same flow. This latter case

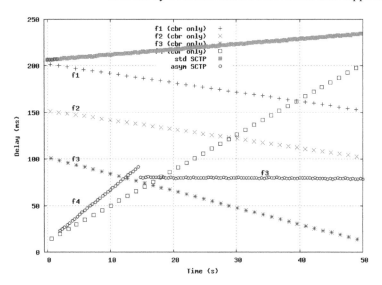

Figure 5.16 Comparison between: (1) Standard SCTP (primary path only: f_1); (2) Symmetric delay-centric SCTP (primary round-trip path ($f_1 r_1$) has the lowest RTT; (3) Asymmetric SCTP (paths f_1, f_4, f_3); (4) Delay of CBR background cross traffic (with no SCTP traffic) was also plotted for reference.

indicates a pressure for more bandwidth by the current flow. Depending on the traffic characteristics of the flow, the lowest-delay path selection method may also be able to deal with the increase in latency due to additional bandwidth requirements of the flow itself. This would be related to how frequent the delay probes can successfully detect the increase in path latency. Figure 5.17 displays such situation. The SCTP flow has its data rate doubled to 4 kbps. At times t=39, 41, 46 and 48 s the selected path keeps changing between f_3 and f_2. The momentary increase in delay at each of these paths is due to self traffic because otherwise they would exhibit the same background traffic delay pattern previously obtained.

RTT probe interval can be adjusted as a compromise between responsiveness to delay changes on the paths and the overhead traffic generated by the probes themselves. Some optimizations can be done when implementing this method. The current path in use does not need to have HB probes sent on it because SRTT is already being estimated by normal protocol operation based on SACKs received. The alternate paths also have their SRTT updated when an HB-ACK is received. The required modification is to have the HB-ACK also sent by the receiver to its primary destination. Delay monitoring is to be done for every SCTP association. Some implementation may choose not to repeat measurement to same destination address and share the SRTT information among associations with same endpoint (layer 3 address).

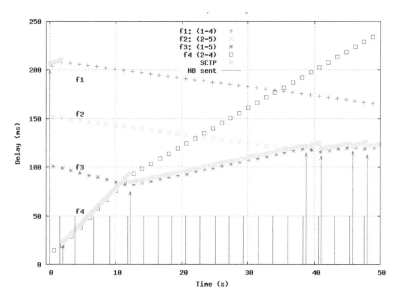

Figure 5.17 Delays on forward path with SCTP bandwidth increase (packet size=100, interval=0.2 s). Vertical lines at the bottom indicate when HB probes were sent. Vertical arrows indicate when transmission is switched to a new path with lower delay. At t=38 s SCTP switches back and forth between paths f_2 and f_4 due to self traffic loading on current selected path.

5.4 Discussions

Multihoming capability of SCTP has established an important framework that is being explored to improve communication performance. Realtime multimedia applications can greatly benefit from such approach. Wireless devices in heterogeneous networks are transforming the classic network scenario. Networks with diverse characteristics in terms of signal propagation, bandwidth, delay and packet loss offer different opportunity for multimedia traffic distribution. The dynamic nature of such environments where those characteristics may change in a matter of seconds bring up the challenge to have a good strategy for selecting the most adequate network for use at a given time.

The simple idea to select the path which has the current lowest end-to-end estimated delay is very promising. It has been demonstrated through simulation and experimental tests that this method provides seamless handover when necessary, granting low overall latency for the realtime communication. Simulations with controlled background traffic have shown how users' perceived quality in terms of MOS calculated by E-model behave as a function of the delay in the network for a VoIP CBR transmission. A simplistic scenario where background traffic was generated by a Markovian process with the same mean packet interval on both paths presented an interesting result. Delay-centric handover method can maintain an elevated quality even for high mean delay path values where a single path transmission would have its perceived quality significantly reduced. This is a good demonstration of the ability of the method to take advantage of the short periods of less traffic on each path. It is expected a similar behavior for real traffic which exhibit higher variability.

The problem of greedy selection of the paths when many simultaneous SCTP sessions employ the same delay-centric method was analyzed. Simulations have shown that SCTP associations can independently distribute their traffic among the available paths. It was verified that oscillations can occur and cause reduction of voice quality. Nevertheless, this unstable behavior only occurred when the utilization factor due to VoIP CBR transmissions was very high (close to unity). It can be argued though that this is not an expected situation because one do not foresee to operate an aggregate of CBR-only traffic close to this limit. On the other hand, conventional single-path transmission would probably not reach an ideal balance among the paths. Any unbalance would cause a significant degradation of the MOS for the path with more VoIP sessions (which would have increasing delays on its queues). Anyway the delay-centric method was not robust enough to deal with this boundary situation gracefully. There is room for some improvement in this sense.

Asymmetric round-trip path method for selecting one-way path with lowest delay is a very interesting approach to further improve the low-delay communication requirement. There is a increase in the number of possible round-trip paths that can be used. There are M^2N more options for a $N \times M$ multihomed system when compared to symmetric round-trip paths. The lowest one-way path can be used for the multimedia transmission in each direction. Examples illustrated the case when a forward path has the lowest latency but its corresponding return path has a high latency. A symmetric round-trip path method does not select this path for transmission. Asymmetric method resolve this problem as it can compares one-way delays.

Simulations on NS2 demonstrated the operation of the asymmetric method in a scenario where a variable number of TCP background traffic induces delay on both one-way paths differently. The modified SCTP implementation was able to compare SRTTs for each asymmetric round-trip path and determine the lowest one-way path.

For a successful wide deployment of low delay SCTP communication further investigations should be carried on. How the loss rate would interfere with path selection mechanism is certainly an important topic specially for wireless networks. The interaction between delay-centric selection and SCTP failover mechanism is another point that needs further exploration. For a mobile device it will be useful to combine the path selection method with other layer-2 information like RSS (Received Signal Strength). So far many investigations have addressed those situations separately but few have studied them together. Also the fine-tune of important parameters like HB rate, delay hysteresis and SRTT calculation should be further examined.

Bibliography

Barakat C, Altman E and Dabbous W 2000 On TCP performance in a heterogeneous network: A survey. *IEEE Communications Magazine* **38**(1), 40–46.

Budzisz L, Ferrús R, Grinnemo KJ, Brunstrom A and Casadevall F 2007 An analytical estimation of the failover time in sctp multihoming scenarios. *WCNC 2007 proceedings* pp. 1–6.

Caro A and Iyengar J 2005 SCTP module for NS2. *http://pel.cis.udel.edu*.

Caro Jr. A, Amer P and Stewart R 2004 End-to-end failover thresholds for transport layer multihoming *Military Communications Conference*, vol. 1, pp. 99–105 Vol. 1 IEEE.

Caro Jr. AL, Amer PD and Stewart RR 2006 Retransmission policies for multihomed transport protocols. *Computer Communications* **29**(10), 1798–1810.

Caro Jr. AL, Armer P, Conrad P and Gerard Heinz GJ 2001 Improving multimedia performance over lossy networks via SCTP. pp. 1–5.

Fiore M and Casetti C 2005 An adaptive transport protocol for balanced multihoming of real-time traffic. *Global Telecommunications Conference, 2005. GLOBECOM '05. IEEE*, vol. 2, pp. 1091–1096.

Fitzpatrick J, Murphy S and Murphy J 2006 SCTP based handover mechanism for VoIP over ieee 802.11b wireless lAN with heterogeneous transmission rates. *IEEE International Conference on Communications - ICC'06*. pp. 1–6.

Fitzpatrick J, Murphy S, Atiauzzaman M and Murphy J 2009 Using cross-layer metrics to improve the performance of end-to-end handover mechanisms. *Computer Communication Preprint Online* **10.1016**, 13 pages.

Fitzpatrick J, Murphy S, Atiquzzaman M and Murphy J 2008 Echo: A quality of service based endpoint centric handover scheme for VoIP.

Fracchia R, Casetti C, Chiasserini C and Meo M 2007 Wise: Best-path selection in wireless multihoming environments. *IEEE Transactions on Mobile Computing* **6 (10)**, 1130–1141.

Fu CP and Liew SC 2003 A remedy for performance degradation of TCP vegas in asymmetric networks. *IEEE Communications Letters* **7**(1), 42–44.

Gavriloff I 2009 *Evaluation of delay based path selection mechanism in multihomed systems using SCTP* Master's thesis Federal University of Parana.

ISO/IEC 2008 14496: Coding of audio-visual objects – part 10: Advanced video coding. *Information technology*.

ITU-R 2001 BS.1387: Method for objective measurements of perceived audio quality. *Broadcasting service (sound)*.

ITU-T 1988 G.711: Pulse code modulation (PCM) of voice frequencies. *Transmission systems and media, digital systems and networks*.

ITU-T 1993 H.261: Video codec for audiovisual services at p x 64 kbit/s. *Audiovisual and multimedia systems*.

ITU-T 1996 P.800: Methods for subjective determination of transmission quality. *Methods for objective and subjective assessment of quality*.

ITU-T 1998 P.862: Perceptual evaluation of speech quality (PESQ): An objective method for end-to-end speech quality assessment of narrow-band telephone networks and speech codecs. *Methods for objective and subjective assessment of quality*.

ITU-T 2005 H.263: Video coding for low bit rate communication. *Infrastructure of audiovisual services — Coding of moving video*.

ITU-T 2006 Y.1541: Network performance objectives for IP-based services. *Internet protocol aspects — Quality of service and network performance*.

ITU-T 2008a E–model R–value calculation tool. *http://www.itu.int/ITU-T/studygroups/com12/emodelv1/*.

ITU-T 2008b G.107: The E-model: a computational model for use in transmission planning. *Transmission systems and media, digital systems and networks.*

ITU-T 2008c J.247: Objective perceptual multimedia video quality measurement in the presence of a full reference. *Cable networks and transmission of television, sound programme and other multimedia signals.*

ITU-T 2009 H.264: Advanced video coding for generic audiovisual services. *Audiovisual and multimedia systems.*

Kashihara S, Iida K, Koga H, Kadobayashi Y and Yamaguchi S 2003 End-to-end seamless handover using multi-path transmission *Proc. IWDC*, pp. 174–183.

Kelly A, Muntean G, Perry P and Murphy J 2004 Delay-centric handover in SCTP over WLAN. *Transactions on Automatic Control and Computer Science* **49**(63), 211–216.

Mascolo S, Casetti C, Gerla M, Sanadidi M and Wang R 2001 TCP westwood: Bandwidth estimation for enhanced transport over wireless links. *Proc. of the ACM Mobicom 2001.*

Network Simulator 2 n.d. http://www.isi.edu/nsnam/ns.

Noonan J, Perry P, Murphy S and Murphy J 2004 Simulations of multimedia traffic over SCTP modified for delay-centric handover. *World Wireless Congress.*

Prasad RS, Murray M, Dovrolis C, Claffy K, Prasad R and Georgia CD 2003 Bandwidth estimation: Metrics, measurement techniques, and tools. *IEEE Network* **17**, 27–35.

Ribeiro EP and Leung VCM 2005 Asymmetric path delay optimization in mobile multi-homed SCTP multimedia transport *WMuNeP '05: Proceedings of the 1st ACM workshop on Wireless multimedia networking and performance modeling*, pp. 70–75. ACM, New York, NY, USA.

Ribeiro EP and Leung VCM 2006 Minimum delay path selection in multi-homed systems with path asymmetry. *Communications Letters, IEEE* **10**(3), 135–137.

Santos MN, Ribeiro EP and Lamar MV 2007 Measuring voice quality using a VoIP simulated network. *Proceedings of the international workshop on telecommunications* pp. 137–141.

Stewart R and Xie Q 2001 *Stream control transmission protocol (SCTP): a reference guide.* Addison-Wesley Longman Publishing Co., Inc.

6

High-Performance Computing Using Commodity Hardware and Software

Alan Wagner and Brad Penoff

One application area that has made use of SCTP and its features has been middleware for MPI (Message Passing Interface), a library widely used in High Performance Computing to write message-passing programs for scientific computation. The group at the University of British Columbia has released modules for both major open source MPI libraries: MPICH2 from Argonne National Laboratory and Open MPI from a consortium of academic, governmental, and industrial research labs. The SCTP-based middleware modules make it possible for the wide-variety of currently available MPI programs to execute using SCTP as the underlying communication protocol, without modification to the applications themselves. The middleware can take advantage of many standard features of SCTP including multihoming as well as those in extensions such as the CMT in the FreeBSD implementation.[1] Utilizing these SCTP features, this work has been able to improve the reliability and performance for clusters of commodity machines for use as compute servers. In this chapter, we describe our experiences in using SCTP and multihoming features, and discuss the design as well as some experimental results.

6.1   Introduction

There are several features of SCTP that make it a good choice as a transport protocol for IP-based computing clusters. Support for multihoming is one of the important features of SCTP that can be used to improve the reliability and performance of applications executing inside the cluster. In particular, MPI programs can significantly benefit from multihoming in addition to other features of SCTP. The central focus of this chapter is to use and evaluate SCTP and its features like multihoming in the design of MPI middleware for executing parallel message-passing applications on low cost compute clusters assembled from commodity off-the-shelf parts.

This chapter first introduces MPI and its typical runtime environment in 6.2. An overview of the design of Open MPI as a whole is presented in 6.3 where as the SCTP-based modules we incorporated into Open MPI are introduced after that in 6.4. The chapter concludes in 6.5 where the performance of the system is evaluated.

6.2   Message Passing Interface

6.2.1   Overview

MPI emerged in the early nineties in response to the growing number of platform-specific message-passing libraries that was used for each new architecture. There was a need to form some consensus in order to ensure that application developers did not continually have to port their code. MPI has been enormously successful in this regard; there are MPI-based applications such as PETSc (Balay et al. 2009) for the parallelization of problems modeled as partial differential equations, scalable libraries for Eigenvalue problems, computational fluid dynamics (CFD) solvers, and plug-ins for many basic tools like Matlab. The standardization process for MPI was done by a consortium of industry, academic, and application specialists who ensured that the library was sufficiently expressive to take advantage of the many styles of message-passing yet capable of executing efficiently on a variety of parallel machines ranging from tightly coupled shared-memory architectures to loosely coupled distributed memory systems in local and even wide-area networks. MPI has continued to evolve and MPI 2.0, standardized in 2002, is now widely available. The MPI

---

[1] Concurrent Multipath Transfer was introduced in Chapter 4.

forum has re-convened for an MPI 3.0 that was standardized in 2012. Although MPI was originally intended for high performance machines, today with multicore and clusters being so commonplace, MPI is also being using in smaller sized systems.

Clusters computers have become easier to assemble and manage over the last decade. Although most major computer vendors provide tools for cluster management and in many cases their own version of the MPI libraries, there are many mature open-source software toolkits available for installing and managing small to medium sized clusters consisting of some number of low-cost computers and switches: *Ganglia Monitoring System webpage* (2009); Rocks Cluster Toolkit webpage (2009). In the case of MPI, for IP-based clusters, there is Open-MPI (2009) and MPICH2 (2009), which are open-source versions of the MPI library; these libraries continue to support the latest architectures and network fabrics. In smaller settings, these libraries are often used in IP-based networks.

MPI was originally designed for tightly coupled applications that execute on fast, reliable systems. As such, it has very limited support for reliability and the usual distributed systems concerns with respect to failure. The typical MPI program executes as a single program and failure of any process, and in many cases the network, causes the entire program to terminate by default. There has been efforts to enhance the fault tolerance of MPI through proposed changes to the library (Graham et al. 2002). In addition, fault tolerance was an area addressed in MPI 3.0.

For IP-based networks, major concerns include fault tolerance as well as performance. Fault tolerance relies on IP's best-effort semantics and network conditions can cause paths in the network to fail. This is even more the case on systems that may depend on low-cost network switches which can show anomalous behavior. The performance of IP networks is also an issue because of the relatively high cost of communication. The cost of communication can limit the amount of parallelism that can be effectively exploited in the system. In IP-networks, the copying and protocol processing overheads make it difficult to overlap communication with computation in local area networks since the majority of the transfer time occurs at the end-stations. As a result, bandwidth can dominate message latency making it all the more critical to provide applications with as much bandwidth in the underlying system as possible. SCTP can be used to mitigate both fault tolerance and performance factors in IP-based clusters.

The first step in using SCTP for MPI is to develop middleware components that use SCTP. In Section 6.2.2, we introduce the general structure of an MPI implementation, then in Section 6.3, we introduce the general design of Open MPI before describing in Section 6.4 our Open MPI SCTP-based modules and eventually evaluating them in Section 6.5.

6.2.2 MPI Runtime Environment

An MPI installation consists of more than simply a library for the MPI API itself. Figure 6.1 shows the basic structure of an MPI application and the middleware required to run it on a typical system.

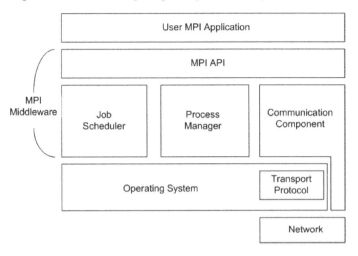

Figure 6.1 Components of an MPI program and implementation.

At the top, there is the application program and at the bottom there is the actual resources that make up the runtime system, for example an operating system and the underlying hardware. MPI middleware is responsible for glueing these two parts together and maintaining the appropriate state all the while. MPI was defined independent of the underlying runtime system of the machine. Generally, MPI requires some type of external scheduler to start and terminate processes; this is what is called the job scheduler. Once a process exists, a process manager is then necessary to pass signals/IO to/from the active processes. These components can vary as the runtime system can be a queuing system as found in large clusters or a standalone process manager. Throughout these allocated processes, the application program executes as a collection of processes on the target machine(s). Because MPI supports both blocking and non-blocking communication, there needs to be some middleware layer component that manages state associated with the delivery of messages between processes; this is the communication component in Figure 6.1. The middleware layer hides the underlying transport mechanism from the application programmer and, in this manner, MPI can support a variety of dedicated fabrics as well as standard IP transport protocols. Adding a transport mechanism to MPI requires adding and configuring new communication modules to the middleware of MPI.

The MPI-2 standard contains over 200 API calls supporting a variety of communication primitives including the basic send/receive operations which can operate either synchronously or asynchronously. MPI allows asynchronous communication via non-blocking primitives to overlap computation with communication, hence helping to improve the latency tolerance of the application. Aside from point-to-point communication, MPI also provides mechanisms for collective communication that allow for the creation of a group of any subset of MPI processes, and then to perform communications amongst that group such as broadcast, reduce, gather, and all-to-all.

On start-up, each MPI process is assigned to a globally unique rank (from 0 to N-1, where N is the size of the MPI universe) and the rank is used to address an MPI peer with respect to the participants of the parallel application. A typical MPI message has an envelope attached that contains the metadata of the message; the metadata includes the type of message, source and target rank, and a tag value. In addition, for point-to-point operations, each message is also designated by an opaque value that serves an MPI-level context ID which is used to separate collections of MPI messages. In particular, the context ID is used to differentiate MPI primitives from each other. For instance, MPI collective messages should not be mixed with simple point-to-point communication. By including a context value, MPI middleware can ensure that MPI messages from different primitives are matched correctly. Message matching is performed at the TRC level (Tag, Rank, and Context) and not every operation requires the use of a tag (e.g., collective operations).

In addition, MPI allows the use of wildcards in posting a receive call to match from any source (MPI_ANY_SOURCE) and/or any tag (MPI_ANY_TAG). MPI also provides the abstraction for process groups and communicators that allow the user application to parallelize their problem domain in a more descriptive way. Overall, the MPI specification exports a comprehensive set of abstractions to the users and by using a common specification; this can improve the portability of MPI programs for different version of the MPI middleware. Internally, MPI defines the notions of short and long messages as well as expected and unexpected messages. Short messages are transmitted immediately while long messages (larger than a threshold value) require the use of a rendezvous protocol to prevent the sending side from overflowing the receiver's buffer. An expected message is one for which the receiver is waiting ("posted") and an unexpected message is an MPI message that was not able to be matched and thus will be buffered.

We have designed several versions of MPI middleware for MPICH2 (2009) and Open MPI (2009). The middleware uses SCTP as the transport protocol and takes advantage of many of the features of the protocol. In the next section, we introduce the general design of Open MPI middleware. Following that section, the SCTP-based modules we designed for the Open MPI middleware are described.

6.3 Open MPI Communication Stack

Open-MPI is an open-source implementation of MPI-2. The Open MPI consortium responsible for the creation of Open MPI has representation from industry, academia and US national lab (Open MPI, 2009) and builds upon work from previous open-source implementations of MPI. Open-MPI is built around the concept of a component architecture that provides services to manage different component frameworks to allow users to mix-and-match these frameworks to instantiate a customized version of the MPI middleware and runtime system. Open MPI has different components to support different network fabrics and the component architecture also makes it easier to simultaneously activate components making it possible to use multiple fabrics such as InfiniBand and Ethernet at the same time. In our design of SCTP-based modules for Open MPI, rather than design our own framework we started with a duplicate of the framework used with TCP.

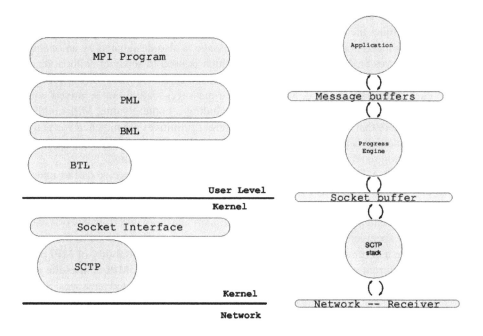

Figure 6.2 Architecture of the Open MPI system.

A functional diagram of the different layers to support messaging on top of SCTP is shown in Figure 6.2. The top layer on the left is the application program that is periodically calling MPI functions to pass messages between processes. The PML layer (Point-to-point Messaging Layer) maintains most of the message state and manages the sending and final delivery of messages to the appropriate receive call and user buffer in the program. Directly below the protocol neutral PML layer there is the BML (Byte Management Layer) that is a relatively thin layer that is used, in the case of TCP, to manage the opening and closing of sockets. The BML layer is also capable of managing several open connections at the same time for a given protocol. Finally beneath the BML is the BTL (Byte Transfer layer) which is the module responsible for moving bytes over the wire or in the case of TCP the module that makes the read and write socket calls. In the case of SCTP, it is the BTL layer which needs to be customized to support SCTP functionality. The BTL layer uses the standard socket API to copy data into the socket buffers where the SCTP protocol stack sends the data on the designated association.

On the right in Figure 6.2 we have illustrated the event-driven relationship between the application, progress engine in the PML, and finally the protocol stack. The circles represent the state machines that either poll or are activated by call-backs to manage the passing of data down to the network. The application and the middleware execute as part of the same process and control is based back and forth between the application and the progress engine by explicit calls to MPI. There is a similar call and call-back relationship between the progress engine and functions inside the BTL layer. Once the BTL layer is called, it attempts to drain the socket buffer and/or write data into the designated socket. Finally, inside the protocol stack itself, there is the state machine to handle queuing and buffering of messages on streams inside SCTP as well as the standard mechanisms for managing the congestion windows.

The message progression engine, or progress engine manages the lifecycle of an MPI message which includes, matching a message receive with the appropriate message and managing the eager sending of small messages versus rendezvous for longer MPI messages. The progress engine polls the underlying communication channels on every MPI library call attempting to progress messages by either sending data for messages that are pending or receiving data for messages currently in the channel. For example, for the standard socket interface, the progress engine manages the select call for the different opened connections of the different MPI processes. We are assuming a passive progress engine that executes only during MPI calls. There are versions of MPI that have explored the use of using separate threads to asynchronously progress messages (Aumage et al. 2007).

In the next section we describe two separate designs that use SCTP with Open MPI. We begin by describing a one-to-one model that is based off of the BTL of TCP. The major difference between the TCP and our SCTP BTL is that SCTP is message based and not stream based. We then describe our implementation of a one-to-many design which makes use of a variety of SCTP features and differs substantially from the TCP module. One difference is that unlike the one-to-one implementation only one socket is opened similar to UDP. This affects the calls between PML and BTL since now messages to and from all processes are received into the same socket. In addition we also investigate the uses of multi-streaming and discuss how multihoming is supported in the two implementations.

6.4 Design of SCTP-Based MPI Middleware for Open MPI

The design of the SCTP modules was based on the TCP module, but with several important differences. In order to support the use of multiple network interfaces there are two possible approaches. One approach is to make use of Open MPI's ability to open multiple connections and the second approach is to allow SCTP itself to handle the multiple connections whereby, with respect to Open MPI, SCTP is viewed as one connection. In the first case we implemented a One-to-One-based version of the software.

An important advantage of using SCTP rather than TCP is that SCTP can support multiple associations within a single connection.

6.4.1   One-to-One SCTP BTL Design

The one-to-one style SCTP BTL is similar to the TCP design and thus the code modification was kept to a minimum. Connection setup and maintenance is also similar to the TCP, even though we are using SCTP sockets.

Connection Establishment

Prior to any message exchange between processes, connection establishment must take place. We will discuss this procedure by utilizing the simple example of 2 processes that want to communicate. Process 1 will be sending a message, in our case a simple integer (though the data type is of little importance in this discussion) to process 2. Given the small size of the message (under 64K), Open MPI will use eager sends. We will later discuss the implication of sending messages over 64K in size.

The first time our BTL code is executed on the sending side, it is invoked from within the Byte Management Layer (BML). The BML triggers a `btl_send` function pointer that was set up by the PML. In our case, this pointer redirects us to `mca_btl_sctp_send()`. Here some interesting things happen. Firstly, we allocate a `mca_btl_sctp_frag_t*` frag structure (from now on known only as `frag` and associate with it the endpoint to which we will be sending our data. This allows us to access addressing information from the `frag` itself which comes in quite handly later on. The `frag` is then packed with header information, the data we wish to send, and has other fields such as `iov` (iovector) counters and indexes initialized. The iovector counters allow us to keep track of how many blocks of data there are to send using our vector write function. Once initialized and packed, the `frag` is handed off to `mca_btl_sctp_endpoint_send()` located in `btl_sctp_endpoint.c`.

The endpoint-to-endpoint send routines are heavily dependent, in the one-to-one BTL version, on connection state. Each `btl_endpoint` data structure that a process maintains has information about its connection state (stored as `endpoint_state`) with that particular endpoint. This connection state is also maintained if the processes are located on the same node. The states maintained are as follows:

1. MCA_BTL_SCTP_CLOSED

2. MCA_BTL_SCTP_CONNECTING

3. MCA_BTL_SCTP_CONNECT_ACK

4. MCA_BTL_SCTP_FAILED

5. MCA_BTL_SCTP_SHUTDOWN

6. MCA_BTL_SCTP_CONNECTED

On our initial send, the state of the connection that the sender maintains for the node to which it is sending is MCA_BTL_SCTP_CLOSED. In other words, this is the first time we have tried to send anything to this particular process, and thus no connection exists between us. In order to transfer our message, we must first enter a fairly involved connection establishment phase, during which some application level handshaking occurs. This phase is initiated by calling mca_btl_sctp_endpoint_start_connect().

Each btl_endpoint structure has an endpoint_sd field used to store a socket file descriptor. Upon initiating connection setup, we assign a new SCTP streaming socket to this endpoint_sd field. We then bind() a NIC to the newly created socket, set the socket to non-blocking, setup event callbacks, pull the addressing information from btl_endpoint into a sockaddr_in struct, and finally we call connect() on this new socket. If our connection does not succeed immediately, we wait for completion by adding a send event to an event queue maintained by Open MPIs upper layers. In other words, we notify Open MPI's progress engine that we are waiting for a send event to complete. We then update endpoint_state to MCA_BTL_SCTP_CONNECTING.

Assuming that we eventually succeed in our connection attempt, we are obligated to send our (i.e., the processes) unique, global identifier (guid) to the receiving endpoint. This identifier consists of our virtual process identifier (vpid), which is used to identify processes in MPI-based programs, and a jobid. The guid itself is stored within the processes local mca_btl_sctp_proc_t* btl_proc struct under the proc_name field. Note that a process can access its local btl_proc structure by calling mca_btl_sctp_proc_local(). The actual sending of the guid is performed via a blocking send to ensure the guid arrives before we attempt to send any other data. mca_btl_sctp_endpoint_send_blocking() is called and the guid is sent on the socket created above using the sctp_sendmsg() call. If the guid was sent correctly, then the endpoint_state is updated to MCA_BTL_SCTP_CONNECT_ACK to reflect this. At this point, the local process will add a receive event to the event queues, signifying that we are now waiting for a response from the process to which we will be sending data. This is all part of the application level acknowledgment scheme employed by Open MPI. We are then pushed out into the Open MPI's progress engine where we spin until there is something more for us to do.

We now switch to the second process which will be the receiving process in our example. The sender is currently spinning in the progress engine, waiting on some kind of acknowledgement from the receiver. Much like with the sender, there is a fair bit of setup that occurs when the MPI program makes a receive call. Resources, such as buffer space are allocated and receive requests are initialized and posted. Once the progress engine is aware that there is a connection waiting for the receiver it triggers the receiver's mca_btl_sctp_component_recv_handler(). The first thing the component level receive handling code does is check which socket the data came in on. If it's directed at the receiver's listening socket, then it's a connection request arriving and so the mca_btl_sctp_component_accept() is called to accept the incoming connection. Once accept() has been called, we are put back into the component level receive handler where we call sctp_recvmsg() in order to pull the sender's guid off the socket. We then determine if the guid belongs to a valid process and check if the connection request matches our endpoint's address. Assuming that it does, we perform a check to ensure we do indeed wish to communicate with this process and finally send them our local guid. If this send succeeds, mca_btl_sctp_endpoint_connected() is called which finally sets the endpoint_state to MCA_BTL_SCTP_CONNECTED. Please note that a similar guid receive session occurs at the sender once the receiver has sent it's guid as part of the application level hand-shaking procedure. This completes the connection setup phase for the one-to-one style SCTP BTL. We are thrown back into the progress engine and are now ready for application level data exchange.

Sending Application Data

Assuming all went well during connection setup, our endpoints are now connected and the sending process can begin to send data to the receiving process. The progress engine gets notification that the sender now wishes to send it's application level data. We hit the callback for this event and are dropped into the mca_btl_sctp_send_handler() routine, which is responsible for completing the current send. After checking the state of the connection, it initiates the fragment send code located in mca_btl_sctp_frag_send(). The fragment send code does a little more than just call our vector writer. The first thing it does is determine if the message we are trying to send is classified as a large message. The way that it does this is by simply checking each iovector structure belonging to the particular fragment to determine if the cumulative length of all the *iovectors* exceeds the maximum message size of 64K. I will discuss the procedure of sending large messages later in the document. For the time being, we determine the message we wish to send is not large enough to warrant any special treatment and so we simply call our vector writer which starts to send out the fragment iovectors containing both the application level message as well as MPI header information. This header information allows the upper levels to match and deliver the message to the appropriate process. Note that these writes are non-blocking and so can be interrupted. Our vector writer returns how much data was actually sent on the wire and we update the iovector state (i.e., the number of iovectors left to send, the next iovector to send) of the fragment we're sending to ensure completion. If the contents of the entire fragment iovector structure was not sent, we will return to try again after a spin in the progress engine.

Once the send has completed, the sender gets thrown back into the progress engine where it waits for something, if anything to do. If it has finished doing what it set out to do (in our case it has), the progress engine releases the sender and it returns, eventually falling out into main() routine of the MPI test program.

Receiving Application Data

When the progress engine for the receiving process gets notified of data waiting on the socket, it drops us into mca_btl_sctp_endpoint_recv_handler(). This routine checks the state of the connection between us and the sending process and according to this state, takes an appropriate course of action. The state that we are currently in is MCA_BTL_SCTP_CONNECTED so we know that the data to pull off the socket is application level data. We allocate a fragment pointer and assign it the fragment pointer contained within the btl_endpoint structure that is our view of the endpoint with which we are communicating (in this case, the sending endpoint). We pass this fragment to mca_btl_sctp_frag_recv() which is actually responsible for pulling the data off the socket. Once inside, this routine is essentially responsible for packing data pulled off the socket via readv() into the fragment iovector structure. Like the write, this read is also non-blocking and can be interrupted, as a result, bookkeeping operations such as iovector pointer and counter updates are performed to ensure that we accurately collect all the data bound for this particular fragment.

Once all the application level data has arrived and the fragment has been packed appropriately, we return true to the endpont receive handling routine which signals that we are done reading and that the fragment is ready to be shipped off to the upper layers of Open MPI where it can be delivered to the user MPI program.

6.4.2 One-to-Many SCTP BTL Design

This was the second phase of the SCTP BTL project. We move away from using a TCP-like streaming SCTP socket to a UDP-like message oriented SCTP socket. We take advantage of the fact that we can now maintain only one active socket and use this socket to send and receive all data. As a result, this implementation scales well as cluster size increases.

There were a number of changes to the design that had to be made to allow for the use of only one socket for all incoming and outgoing communication. The changes and implications are detailed below.

Socket Changes

First of all, the socket itself had to be changed from a SCTP SOCK_STREAM socket to a SCTP SOCK_SEQPACKET socket.

Connection States

The one-to-one SCTP BTL, like the TCP BTL, uses connection states between endpoints to determine the correct course of action to take during a send or receive cycle. For example, if the endpoints are not connected, then guid exchange occurs. If, on the other hand, the connection has been established and the application level handshake has occurred, then the endpoints will actively send or receive application level data.

For the one-to-many implementation, because we only have one socket to work with, we rely on the idea of implicit connect and have completely removed any connection state functionality found in the one-to-one BTL. Implicit connect simply means that any time a process sends data to another process, those two endpoints form a very lose connection or association (not to be confused with a SCTP association). In effect, the two endpoints become aware of each others existence, via hashed addressing information, which allows them to send and receive data from one another. This also means that we now have to keep track of endpoints with which we have an implicit connection. This is done by using hash tables to store data structures that identify endpoints with which the process has communicated with in the past.

Application Level Handshake

We saw in the previous section that Open MPI performs an application level handshake (i.e., `guid` exchange) before any user MPI program data are exchanged. In the one-to-many version, we felt the need to modify how this handshake is performed. Normally, the sending process will, assuming the connection state is not `MCA_BTL_SCTP_CONNECTED`, send its `guid` to the receiver and the receiver will mirror this operation by sending it's `guid` back to the sender. In the one-to-many case, we delay the receiver from sending it's `guid` back. Instead, it simply uses the `guid` it received from the sender to reference it's hash table and insert an entry for the endpoint if needed. It can then begin receiving data from the sender. If, however, the receiver wishes to send something back to the sender, it will reference another hash table to see if it has ever sent anything to that particular endpoint. If it has not, then it will add an entry for that endpoint in this hash table and will start the transmission by sending it's `guid` followed by any messages. If an entry for that endpoint already exists, then the receiver knows it has sent something to that endpoint in the past and will skip the `guid` send and begin sending data immediately.

Data Buffering

Given that we only use one socket for all sends and receives that a process makes, we run into a serious problem not seen in the one-to-one BTL. All incoming data, from any process, on any node, is sent to that process's single listening socket. As a result, we can not pull the data directly into the fragment structure (as was done in the one-to-one BTL) because we simply can not determine what endpoint is sending the data prior to pulling it off the socket. In the one-to-one case, each connection between two endpoints was unique (identified by a unique socket file descriptor), thus we knew which endpoint was sending us data based on what socket the data arrived on. We could then simply pack the data into the correct fragment. In the one-to-many case, data from numerous endpoints can arrive on the socket and so we need to store data from each read in an intermediary buffer, then determine where the data originated from, locate the correct fragment structure (by checking and/or updating our endpoint hash tables) and finally pack the data. As a result, we decided to amalgamate both the component level and the endpoint level receive handlers into one generic receive handler located in `mca_btl_sctp_recv_handler.c`. The `mca_btl_sctp_recv_handler()` routine is responsible for much of this functionality. It pulls data off the listening socket, buffers it, finds the correct fragment structure and directs both the buffered data and the fragment where it will be stored to the fragment receive code. Note also that in the one-to-many case, the fragment receive code was modified and does not actually perform any socket reads. Instead, it simply packs the fragment with the data it received from `mca_btl_sctp_recv_handler()`.

Multihomed Hosts

The idea of having a single socket also impacts how we utilize more than one link. In the one-to-many case, we make use of Concurrent Multipath Transfer (CMT) (Iyengar et al. 2006) in the transport protocol to stripe messages across two links. Unlike the one-to-one version, we are unable to run more than one one-to-many BTL because Open MPI only exports one listening socket regardless of how many BTLs you intend on using. Therefore, we can only register a single listening socket for performing all communication operations. We get around this problem by using `sctp_bindx()` to bind the second network interface to the single listening socket. This, combined with CMT gives us the ability to utilize both links when sending messages.

Streams

SCTP, aside from associations between hosts, also provides the notion of streams within an association. Ordering of messages is maintained within a particular stream, but not across streams. In other words, if you send several messages on the same stream, they are guaranteed to arrive at the receiver in the same order they were sent. However, if you spread the messages across multiple streams, then the messages may arrive in any order with respect to the streams. This feature allows messages that are independent of each other to arrive without any ordering constraints between them and, if used correctly, this can eliminate TCP's head-of-line blocking (Kamal et al. 2005).

   We wanted to make use of streams in our SCTP BTL, by scheduling a message on a specific stream number based on several MPI parameters (tag, context, and process ID) associated with the MPI-level message. This proved to be a larger problem than orginally considered. Open MPI imposes a strict hierarchy between the various levels (PML, BML, BTL) of its implementation. Their strategy is such that the BTLs are to remain simple byte movers with no knowledge or interest in any MPI semantics. This disassociates the BTL from any other Open MPI layer and allows the BTL to continue functioning should any major changes or replacements occur at the PML or BML level.

   While we found a way to pull the necessary MPI parameters from the fragment structures and our vector writer is capable of sending data on a specified stream number, these ideas would have tied our BTL to the current PML implementation, known as OB1. Since this would break the encapsulation imposed on us, we currently send all our messages on stream 0. As a result, no multiplexing on streams was possible in this SCTP BTL implementation.

Large Messages

As was mentioned earlier, the sending of large messages is handled slightly differently. The check to see if a message classifies as large is performed in the `mca_btl_sctp_frag_send()` function where the size of each of the iovector structures is examined. If the cumulative size is greater than 64K, we classify the message as large and pass the fragment off to `mca_btl_sctp_large_frag_send()`. It is from this function that we perform our vector writes, limiting the size of each attempted send to 64K and updating the iovector pointers to reflect the amount of data written. This is done because SCTP imposes a maximum message size that it can transfer with each send call. Since MPI messages come in varying sizes, we have to ensure, at this level, that any data passed to our vector write code does not exceed SCTP's maximum allowable send size. Since our sockets are non-blocking, we have to perform some additional book keeping to make sure we know exactly where in the fragment our send was interrupted.

6.5 Performance of the System

6.5.1 Reliability

Reliability in communication is important for MPI applications since a single fault anywhere in the communication can result in a complete failure of the computation. There are a variety of ways in which reliability has been addressed in MPI. One technique is to add fault tolerance to software for recovery from errors and it the extreme the application level or system level check-pointing of the application. Although it is difficult to recover from node failure it is somewhat easier to either ensure low failure rates for the network or to attempt to recover when a connection is dropped. Recovering from a dropped connection is easier in an IP-based transport protocol because of the "best-effort" semantics of the mechanisms. SCTP already contains many of the same mechanisms in TCP for congestion control and control flow between end-points. However, SCTP's multihoming adds a new mechanism to use for further improving the reliability of the connection.

Multihoming can increase fault tolerance of applications by automatically failing over and continuing to operate. This is done at the transport layer and is completely transparent to the application. As long as there is path between the IP addresses in the association SCTP the association will continue to operate. The degree of fault tolerance depends on whether or not the paths are independent. Of course, if there is a shared link along the path and that path goes down, then packets can no longer be delivered. Inside a cluster, the cluster designer can build the system to add path redundancy and also ensure the degree of overlap. There is a lot of flexibility in design of fault tolerance. For example, it is possible to optimize the design by making sure that paths across less reliable hardware are made independent whereas paths may share backbone type links across more reliable hardware.

6.5.2 Extra Bandwidth

The performance of MPI programs is often dependent on the latency and bandwidth of the underlying network. In the case of IP-based protocols there in a LAN, when there is no congestion, most of the message time due to bandwidth. In experimenting with the different SCTP BTLs, the focus was on bandwidth. In addition we also measured the efficiency respect to CPU utilization. For MPI with IP-based protocols, as more bandwidth is used, more CPU cycles are needed for the protocol processing which in turn reduces the cycles available for the application work.

We conducted tests in order to measure the potential of our SCTP-based Open MPI BTLs. All the tests were performed on two 3.2 GHz Pentium-4 processors in an IBM eServer x306 cluster. Each of the two nodes had three GbE network interfaces, two of which were private and on separate VLANs connected to a Baystack 5510 48T switch.

Performance Tests

Performance testing of the one-to-one as well as the one-to-many SCTP BTL involved obtaining bandwidth measurements as well as CPU utilization data for BTL operation. For bandwidth testing, we used the OSU Bandwidth package which generates messages of sizes ranging from 1 byte to 4 MB. It then sends these messages via non-blocking MPI send calls to another process and measures the throughput for each message size. Alongside the bandwidth tests, we measured CPU utilization for both the sender and the receiver using iostat

Not only did we test both BTL versions, we also performed tests using varying MTU size, socket buffer size and number of links. For MTU, we initially ran with the standard 1500 byte MTU size. We didn't see any gains over TCP under these conditions, so we increased the MTU size to 9000 bytes. Using these jumbo frames, our BTL out-performed TCP in all but one of the test scenarios.

Testing our BTL under multihomed conditions was done in two ways. For the one-to-one version, we ran the OSU tests with two SCTP BTLs, identical to how a dual BTL run over TCP works. For the one-to-many version, we used a single SCTP BTL (due to the limitation of listener socket registration mentioned in Section 6.6) but enabled Concurrent Multipath Transfer (CMT) in the FreeBSD kernel and used sctp_bindx() to bind the second interface to our BTL in addtion to the first. In the two BTL case (both TCP and one-to-one SCTP), the transfers take advantage of having two socket buffers. However, when using only one one-to-many BTL with two links, the BTL ends up sharing one socket buffer between both links, rendering our BTL at a disadvantage. To get around this, we found it necessary to increase the socket send and receive buffer sizes for the one-to-many SCTP BTL when using CMT. While the one-to-one two BTL setup failed to beat two TCP BTLs, our one-to-many BTL with CMT enabled showed higher throughput numbers than TCP, showing a clear benefit to use CMT.

With respect to CPU utilization, the only test case that yielded any spare CPU cycles was the one-to-one SCTP BTL using only one link. In that case, we saw CPU idle times as great as 60% with the idlest occurring during sends of large messages. The one-to-many SCTP BTL as well as the TCP BTL used 100% of the CPU, on both the send and receive side, regardless of whether we were using one link or two.

Results

The following graphs show the OSU Bandwidth results for all the scenarios tested. We tested varying the BTL design (SCTP one-to-one, SCTP one-to-many, TCP), the number of network links to use, as well as the MTU. All tests were run with the same (500000 byte) socket buffer size.

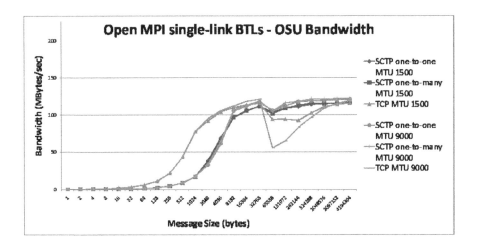

Figure 6.3 Open MPI single-link BTLs — OSU bandwidth.

As is shown, our SCTP BTL outperforms TCP in all but one case. Our one-to-many version is better than TCP in both the single link and dual-link (i.e., with CMT) scenarios. Our one-to-one version beats TCP in the single link case, however, TCP performs better in its two BTL run than our one-to-one two BTL case.

Perhaps the most striking thing in this graph is the dip that is present, in all the tests, regardless of protocol or any other factors such as MTU, socket buffer size, or number of links used. The location of this dip is indicative of where MPI changes from its eager send protocol (for short messages) to rendezvous sends (for large messages). In this case, it happens at the 64K message size boundary. Also of interest is the recovery of throughput. Both the TCP and SCTP BTLs do eventually recover and continue sending at an acceptable rate. Our SCTP BTL, however, recovers much faster than TCP. Furthermore the dip that SCTP experiences is much smaller thus we do not stall nearly as badly as TCP.

As was mentioned above, all but the single link one-to-one version of our SCTP BTL, saturates the CPU (TCP included). Initially, we thought the BTLs themselves

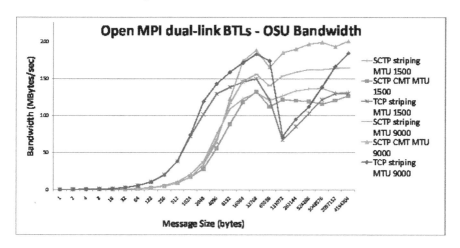

Figure 6.4 Open MPI multi-link BTLs — OSU bandwidth.

were CPU bound, however, this would mean that going from one link to two would produce no improvement in throughput. We showed this not to be the case, which lead us to believe that it was not just the BTL itself that was saturating the CPU, but rather the MPI middleware as a whole.

There is a great deal of time spent within the middleware's progress engine. It is here that socket polling occurs, and send and receive events are dispatched. While inside the middleware's progress function, our user MPI programs cannot perform any work and our BTL cannot send or receive data. We believe that it is the actual progress function that is utilizing the bulk of the CPU cycles. To test this theory, the middleware was instrumented to report how many times we entered the progress engine during each run of the OSU Bandwidth program. The numbers obtained fluctuated between sender and receiver. At times they were similar in magnitude, while other times they were off by an entire order. Given the inconclusive nature of this test, no definitive conclusion was made as to the exact cause of CPU saturation in Open MPI. To further investigate, we turned to the MPICH2 middleware for which an SCTP ch3 device was implemented by several colleagues (Brad Penoff and Mike Tsai). After running OSU Bandwidth tests along with iostat using the MPICH2 middleware, we saw not only higher throughput numbers, but also CPU idle times of roughly 10%. This find helps support our theory that our BTL is not CPU bound but is unable to achieve its maximum possible throughput due to the middleware's use of the CPU.

Bibliography

Aumage O, Brunet E, Namyst R and Furmento N 2007 NewMadeleine: A Fast Communication Scheduling Engine for High Performance Networks *Communication Architecture for Clusters Workshop (CAC 2007), workshop held in conjunction with IPDPS 2007.*

Balay S, Buschelman K, Gropp WD, Kaushik D, Knepley MG, McInnes LC, Smith BF and Zhang H 2009 PETSc webpage. http://www.mcs.anl.gov/petsc/.

Ganglia Monitoring System webpage 2009. http://ganglia.sourceforge.net/.

Graham RL, Choi SE, Daniel DJ, Desai NN, Minnich RG, Rasmussen CE, Risinger LD and Sukalski MW 2002 A network-failure-tolerant message-passing system for terascale clusters *ICS '02: Proceedings of the 16th international conference on Supercomputing*, pp. 77–83. ACM, New York, NY, USA.

Iyengar J, Amer P and Stewart R 2006 Concurrent Multipath Transfer Using SCTP Multihoming Over Independent End-to-End Paths. *IEEE/ACM Transactions on Networking* **14**(5), 951–964.

Kamal H, Penoff B and Wagner A 2005 SCTP-based middleware for MPI in wide-area networks *3rd Annual Conf. on Communication Networks and Services Research (CNSR2005)*, pp. 157–162. IEEE Computer Society, Halifax.

MPICH2 webpage 2009. http://www.mcs.anl.gov/research/projects/mpich2/.

Open MPI webpage 2009. http://www.open-mpi.org/.

Rocks Cluster Toolkit webpage 2009. http://www.rocksclusters.org/.

7

SCTP Support in the INET Framework and Its Analysis in the Wireshark Packet Analyzer

Irene Rüngeler and Michael Tüxen

This chapter describes how the INET framework for the OMNeT++ simulation kernel can be used to simulate IP-based communications and SCTP-based ones in particular.

It is shown how to configure multihomed scenarios and a detailed description of implemented features is given. This allows users to conduct experiments by varying parameters. An architectural overview of the SCTP implementation allows users to extend the provided functionality.

Finally it is shown how simulated nodes can interact with real nodes and Wireshark can be used to analyze SCTP trace files.

7.1   Introduction

The OMNeT++ discrete event simulation environment (an overview is given in Varga and Hornig (2008)) is a very versatile tool, that supports the generation of simulation models by its modular component-based architecture. The INET framework (Varga 2011) is one publicly available model and is very well suited for the simulation of IP-based networks since models for the standard protocols of the TCP/IP protocol family are already provided.

Therefore, the implementation of SCTP formed an ideal extension to the present framework. This SCTP model is now included in the official INET framework.

Since standard conformance was one of the major goals for the SCTP model, it was highly desirable to be able to use the real traffic traces, the test suite implementations for testing real SCTP implementations and of course the readily available real SCTP implementations for debugging, testing and validating the SCTP model. As a consequence, we not only added an RFC conformant SCTP simulation with suitable user applications, but also a dump module to provide traffic traces and an external interface for testing and evaluating the simulation with real-time implementations.

In the next section we will give an introduction to OMNeT++ and INET. The integration of SCTP in the framework focussing on the simulation architecture, the additional modules, the configurable parameters, and the testing is explained in detail in Section 7.3. After two examples showing how to configure multihomed hosts and devices with external interfaces, we will conclude this chapter with the graphical analysis of SCTP trace files in the Wireshark network packet analyzer.

7.2   The Simulation Environment

7.2.1   An Overview of OMNeT++

OMNeT++ is an open source discrete event simulation environment with a modular component-based architecture written in C++ available from Varga (2010). Types of components are channels (described by the parameters delay, bit error or packet error rate and data rate), network definitions, simple and compound modules. The components can be assembled into more complex modules via connected gates. Networks are the result of combined module types that communicate through messages.

One message can be encapsulated into another one, thus being able to simulate the transmission of information via layered protocol stacks. The events are ordered ac-

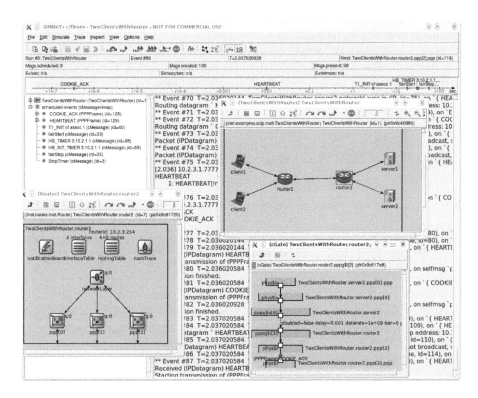

Figure 7.1 OMNeT++ simulation environment.

cording to the time when they should happen using a scheduler. We extended this scheduler to realize the real-time-scheduling needed for the external interface, that we will describe in Section 7.3.3.

The network topology is described using a special high-level language (ned). Parameters can be assigned to modules and easily defined in configuration files.

A powerful GUI helps to follow the simulation process. Each packet is animated and its contents can be shown just by double-clicking on it. Furthermore debug output can be examined for each module individually. Figure 7.1 shows the OMNeT++ GUI. In the left part of the main window the scheduled events are listed, which can be found again in the timeline below the menu. Status information is constantly updated above the timeline. The other windows show only a small collection of the possible 'insides' of the simulated network. The left window shows the compound module `router2` and the right window the gates of the channel between `router2` and `server2`. The channel parameters can be seen and one message just arriving at `router2`. Further information is available by selecting other modules.

Different speed rates for the animation of the simulation can be selected from the toolbar (see Figure 7.1). Every movement of a message can be inspected by stepping through the simulation. The velocities *Run* and *Fast* provide a normal and less detailed animation than the *Step* mode. In the *Express* mode the displayed information is updated only in long intervals.

As it is sometimes necessary to have several runs with just one varying parameter, the application can be compiled to run in the command-line interface. Thus no GUI output will slow down the activity. Since OMNeT++ version 4.0, the simulation

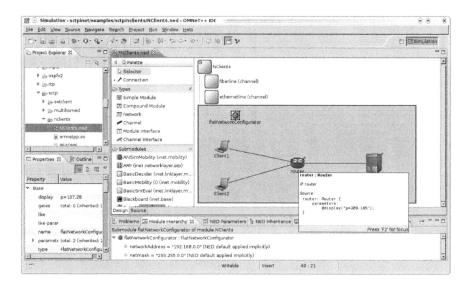

Figure 7.2 OMNeT++ integrated in the Eclipse IDE.

environment is integrated in the Eclipse IDE (see Wilson (2011)). Thus the complete process from building new networks, coding and debugging the sources to analyzing the results is supported in one tool. Figure 7.2 shows the IDE with the tool to set up a new network. Via drag-and-drop submodules can be added and connected with channels, by selecting the possible connection points. Then the properties of the modules can be set, e.g., name, icon, color, size, gates. Simulation results are collected for a later analysis in vector (.vec) or scalar (.sca) files. They can be analyzed in the IDE, too. Figure 7.3 depicts one of the charts generated from one dataset. In the lower window the dataset properties including the number, the mean and the standard deviation are shown.

7.2.2 The INET Framework

As OMNeT++ is a very versatile tool there are a great number of ready-made simulation models provided for download. A popular one of those is the INET framework (Varga 2011).

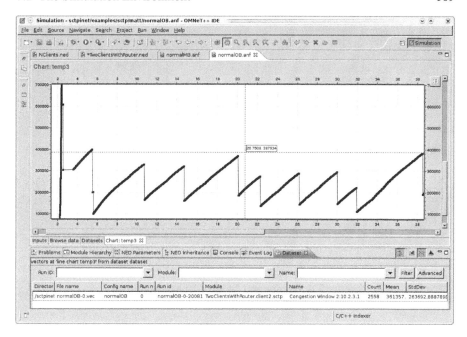

Figure 7.3 Analyzing results of a vector file.

The INET framework is ideal for simulating IP-based networks. The different network layers can be distinguished and layer specific protocols are provided. On the link layer PPP, Ethernet and WLan interfaces can be configured, the network layer features IPv4 and IPv6, routing protocols like OSPF and IP control protocols like ICMPv4, ICMPv6 and RSVP. On the transport layer SCTP, TCP and UDP are implemented. In addition, a lot of protocol independent modules like RoutingTables, Routers, Switches and Hubs are available. They are all implemented as simple modules and can be combined to form compound modules and networks.

One of those compound modules, for instance, is the StandardHost (Figure 7.4) which consists of a complete IP stack with PPP or Ethernet interfaces, a network layer, a Ping application, TCP or UDP as transport layer and corresponding applications. We have complemented this host with the transport protocol SCTP, a suitable application, a dump module and external interfaces. We will specify these modules in the following sections and show configuration examples. Another important feature of INET is the ability to use real network addresses and do the routing according to rules derived from routing tables. Although a FlatNetworkConfigurator can be used to automatically distribute addresses among the hosts of a network, we preferred to set up routing tables where we can configure routes to other hosts or networks. Thus we could arrange networks with several subnets connected via routers.

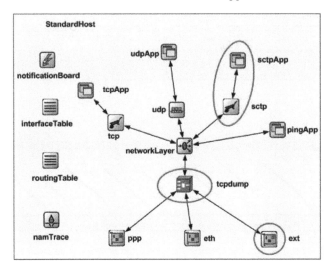

Figure 7.4 Compound Module StandardHost.

7.3   Implementation of SCTP in INET

7.3.1   Realized Features

The INET simulation model has been extended by a fully featured SCTP implementation using the infrastructure of this framework. It supports IPv4 and IPv6 as network layers and multihomed hosts. The base protocol as specified in Stewart (2007) is completely realized. In Stewart et al. (2006) very important modifications concerning congestion and flow control like the calculation of the slow-start threshold, the handling of the congestion window, zero window probing or the sending of gap reports were introduced and also included in the simulation. A more detailed description of the implementation of SCTP in INET can be found in Rüngeler (2009).

One of the main design goals was to be able to freely configure all relevant protocol parameters in order to analyze their influence on the protocol performance.

Looking at Figure 7.4 shows that all this functionality is encapsulated in one simple module called *sctp*. To provide a convenient environment for SCTP protocol and performance analysis a tracer module called *tcpdump* to visualize the message transfers and a traffic generator and collector called *sctpAPP* have been added to the compound module *StandardHost*.

Following the specification of the base protocol, several very useful features have been specified. To be able to change the interfaces used during the lifetime of an association, the Dynamic Address Reconfiguration (AddIP) has been specified in RFC 5061 (see Stewart et al. (2007)). In conjunction with this feature, an authentication is necessary. Thus RFC 4895 (see Tüxen et al. (2007)) was implemented, too. A use case for these features is the Network Address Translation that has been specifi-

cally designed for SCTP in Tüxen et al. (2008b). All these features are integrated in
the simulation and will be made public in a future release of INET.

7.3.2 The Simulation Architecture

SCTP is a complex protocol combining features from TCP and UDP plus realizing
new concepts like streams. Figure 7.5 shows a schematic overview of the different
parts of SCTP. The major blocks that have to be distinguished specify the behavior
of the data sender, the reaction of the data receiver and the control messages including the handshakes to setup and take down an association. Furthermore, Congestion
Control and Flow Control have a significant impact on the behavior of the transport
endpoints. In the following sections we will give a short description of the main parts
of the implementation.

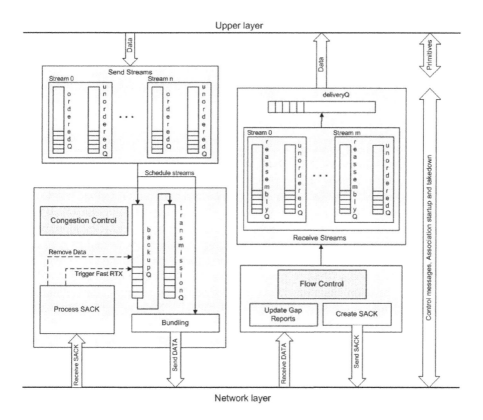

Figure 7.5 The simulation architecture of the SCTP module.

Messages

The means of communication in OMNeT++ are messages. The main class provided is cMessage that defines the necessary attributes and methods to send and receive messages.

The developer can define subclasses by adding message fields. We needed a lot of different classes for the primitives as commands between the application and the transport layer and the messages containing the SCTP packets. The primitives are directly subclassed from cMessage, whereas the SCTP packets are more complex and can consist of several subclasses of *cMessage*. The main classes are SCTPMessage, SCTPChunk and SCTPParameter.

SCTPMessage contains the SCTP message header and a number of chunks. Each chunk type is subclassed from SCTPChunk and has to be handled individually as it has a format of its own. Some chunk types just consist of the header like the COOKIE-ACK chunk, some contain several parameters which can be mandatory or optional. This is the case for the INIT or INIT-ACK chunk. Each parameter has to be subclassed from SCTPParameter, consisting of a header with the type and length information and a body containing the value. When a single message has to be included in another one, encapsulation is used. But when several messages have to be inserted, like the chunks in an SCTPMessage, a dynamic array is used.

Association Setup and Takedown

A four-way-handshake starts an SCTP association. It consists of SCTP-packets containing the control chunks INIT, INIT-ACK, COOKIE-ECHO and COOKIE-ACK and is initiated by sending the primitive SCTP-ASSOCIATE from the upper layer to the transport layer. The receiving side must have sent an SCTP-OPEN-PASSIVE before, so that a listening socket has been created. The handshake is normally started by a client wishing to set up a connection with a server, but a peer-to-peer communication is also possible with both peers opening listening sockets and starting the setup combining their control data to one association. Both alternatives are realized in the simulation and will be referred to in Section 7.3.3. As OMNeT++ is a discrete event simulation environment it uses state machines and provides methods to react to occurring events. In Figure 7.6 the different possible states of the simulation are shown with the corresponding events and the necessary actions to transit from one state to another. The upper half of the figure presents the setup, the lower half the takedown. Data messages are only accepted in the states ESTABLISHED and SHUTDOWN-PENDING.

Timers play a very important role in SCTP. Besides the ones listed in the figure a lot more are needed for instance to trigger the retransmission of data chunks, to send selective acknowledgements (SACKs) or HEARTBEATs. In OMNeT++ timers are realized by sending so-called self-messages. They get a certain arrival time and are inserted in the list of scheduled events. Thus they are handled like other messages and can even carry information.

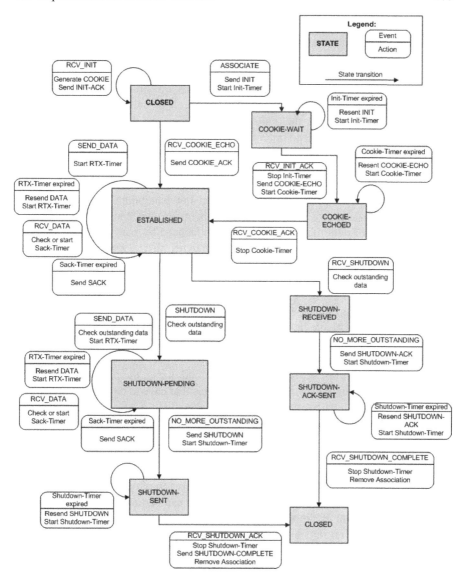

Figure 7.6 Simulation state machine.

Data Sender

After the upper layer received the information that an association had been established it can start sending data (see Figure 7.5). SCTP provides the use of several streams for incoming and outgoing connections the number of which is negotiated in the setup process. Each stream can carry data messages, which can be either unordered or ordered. As the use of the streams is application dependent, the upper layer has to provide the number of streams and also the information which stream each data message belongs to and an indication whether it should be delivered ordered or unordered.

Arriving at the transport layer the data messages are sorted into the appropriate send stream queues. The overall send queue size can be unlimited or limited with configurable size. Whenever the limited queue is emptied to half its size, a notification is sent to the upper layer to order more data.

The sequence in which the stream queues are emptied is controlled by the stream scheduler. The scheduling strategy to be used is not specified by the RFC 4960 (Stewart 2007). Besides the default Round Robin queueing strategy several others, like first-come first-served, priority scheduling, and fair bandwidth have been implemented. They are described in streams and will be made available in a future release.

The amount of data that may be sent is influenced by many factors. It has to be calculated from the number of outstanding bytes, the congestion window, the advertised receiver window and of course the amount of data provided in the send streams. This calculation will be discussed in Section 7.3.2. If the user has configured the sender to consider the Nagle algorithm to reduce the number of small packages (see (Nagle 1984) for details), every packet is bundled with data chunks up to the (configurable) Nagle point. Before inserting them in the packet, a Transmission Sequence Number (TSN) has to be set for the chunk to have a unique means of identification. The data messages that are sent to the peer are stored in the backup queue, until they can be finally removed. A second queue, the transmission queue, is provided for the temporary storage of messages that have to be retransmitted. The information on which path retransmissions occurred, their number, whether the data has been acknowledged or counts as outstanding, and many more attributes characterize a data message and have to be stored with the data. When putting a packet together the messages scheduled for retransmission have to be considered first and only the remaining space can be filled with new data.

An arriving SACK chunk influences the transmission of data as it announces the Cumulative TSN Ack. All TSNs up to this number can be finally removed from the backup queue. Present gap reports lead to an increase in the gap report count of the missing TSNs, which results in a transfer of the chunk from the backup queue to the transmission queue for fast retransmission, if the user defined gap report limit has been exceeded.

Data Receiver

The data receiver is featured on the right hand side of Figure 7.5. The reception of the data messages is influenced by the Flow Control that will be discussed in Section 7.3.2. As the TSNs have to be in sequence, the missing ones are announced in the gap reports that are sent back to the sender for information. The acknowledged data messages are stored in the receive streams. Again each stream consists of two queues, one for unordered and one for ordered data. The order is provided by the Stream Sequence Numbers (SSN) that are maintained for each stream. If data with the appropriate SSN is found, it is stored in the delivery queue, and an SCTP-DATA-ARRIVED-NOTIFICATION is sent to the upper layer, that in turn asks to deliver the data.

The SACK summarizes the results of the TSN analysis and sends the actual Cumulative TSN Ack and the information about the gap reports and possible duplicate TSNs back to the data sender. In addition, the size of the updated Advertised Receiver Window (arwnd) is given (see Section 7.3.2).

Congestion Control

The Congestion Control that SCTP uses is in most parts derived from TCP. Yet some important differences are due to special SCTP features.

As SCTP allows a host to be multihomed, the congestion control mechanism has to be applied to each path separately. This means that a path has its own congestion window (cwnd), slow-start threshold (ssthresh), counter of outstanding bytes and retransmission timeout calculation.

As mentioned before, the Congestion Control influences the amount of data to be sent separately for each path in that not more than the difference between cwnd and the number of outstanding bytes may be transmitted, if permitted by the receiver's arwnd.

While TCP is stream-oriented, SCTP is message based and thus the overhead of many chunks bundled in one packet can lead to a discrepancy between the transmitted bytes and the sent user data. Therefore we have given the user the possibility to switch between the counting of the outstanding bytes including or excluding the header.

Flow Control

The Flow Control as adopted from TCP shall protect a receiver from a fast sender. Therefore, the receiver announces the amount of empty space in the receive buffer by sending the arwnd attribute in the SACK chunks. The simulation follows this approach and reduces the arwnd with every arriving data chunk and increases it when data are delivered to the upper layer. If the window is reduced to zero, the data sender may only send one chunk to probe the window. If a suitable TSN arrives, i.e., one that fills a gap or advances the Cumulative TSN Ack, it has to replace the highest TSN accepted so far, leaving the former unacknowledged again. Announcing this change in the SACK chunk by adjusting the gap reports leads to a change in the attributes of the affected TSNs in the backup queue. Therefore, even TSNs that have been accepted and acknowledged have to be kept in the queue in case they have to be marked as unacknowledged again.

7.3.3   Additional Modules

SCTP Applications

The SCTP module (Figure 7.4) has interfaces to the network layer and the application layer. The interoperability between the transport and its upper layer is realized by sending notifications and primitives (Figure 7.5) as specified in Stewart (2007). In the simulation both a callback and a socket API are realized to provide the upper layer with calls to `bind()`, `listen()`, `connect()`, `send()` or `receive()`. Thus the application layer takes the initiative to start an association. SCTP answers by either sending notifications, indicating for instance that the ESTABLISHED state (Figure 7.6) has been entered, the peer has closed the connection or data are waiting in the receive queue to be picked up.

As examples for upper layer implementations the simulation provides three different applications which work as traffic generators and/or collectors. One is a client with a callback API, one a server with a socket API and the third is a peer that combines both client and server functionality.

The client as a sender can be either configured to send a predefined number of data chunks of a certain length or to start at a certain time and stop at a set time independent from the number of packets. If the packets shall not be sent as fast as possible, a sending interval can be defined. When sending a very large number of messages, the client uses limited send queues, as described in Subsection 7.3.2. The client can also work as a receiver being able to discard or echo the messages.

The server can send or receive data. As it is implemented as a combined server, incoming packets can be discarded or echoed. But the server can also generate data, just like the client, representing a peer-to-peer network. The server keeps a record of all the sent or received amounts of data for each association thus providing statistical data for further use.

To support multihoming, a function `sctp_bindx()` is realized. The user can either set the IP addresses that should be bound explicitly or just leave the default value (empty string), if all available addresses should be used. The bound addresses are included in the INIT or INIT-ACK chunk.

When the peer is initialized, it starts by calling `bind()` and `listen()`, thus being configured as a server. When a certain start time is set, the peer sends a request to SCTP to associate. The peer application is useful, when testing the so-called initialization collisions, that is, when both parties try to set up an association at the same time.

The Dump Module

Although the GUI of OMNeT++ helps to observe the flow of data, we wanted to get a better overview of the packets that were sent to and from the hosts. A dump module was placed between the link layer and the network layer, instead of between the network and the transport layer, to be able to distinguish between the interfaces the message passes through (Figure 7.4). The incoming packets are examined by decapsulating the messages and analyzing their contents. Afterwards they are transferred

```
[2.000] 10.1.1.1.30544 > 10.1.4.1.6666: numberOfChunks=1
INIT
        1: INIT[InitiateTag=267064354; a_rwnd=65535; OS=4; IS=10;
               InitialTSN=1000; Addresses=10.1.1.1,10.2.1.1]

[2.021] 10.1.4.1.6666 > 10.1.1.1.30544: numberOfChunks=1
INIT_ACK
        1: INIT_ACK[InitiateTag=502656; a_rwnd=65535; OS=4; IS=10;
               InitialTSN=2000; CookieLength=0; Addresses=10.1.4.1]

[2.021] 10.1.1.1.30544 > 10.1.4.1.6666: numberOfChunks=1
COOKIE_ECHO
        1: COOKIE_ECHO[CookieLength=0]

[2.042] 10.1.4.1.6666 > 10.1.1.1.30544: numberOfChunks=1
HEARTBEAT
        1: HEARTBEAT[InfoLength=12; time=2.03156]

[2.042] 10.1.4.1.6666 > 10.2.1.1.30544: numberOfChunks=1
HEARTBEAT
        1: HEARTBEAT[InfoLength=12; time=2.03156]
```

Figure 7.7 Output of the dump module.

unchanged to the next layer. Figure 7.7 shows part of the four-way handshake and two HEARTBEAT chunks to different destinations on the server side of the communication. The different IP addresses of the client, to which the HEARTBEATs are sent, are underlined.

The External Interface

During the testing and debugging of the SCTP simulation module it seemed to be very attractive to be able to use existing tools like the Wireshark packet analyzer for analyzing the message transfers and to test the interoperability with existing real implementations. Existing implementations of SCTP test suites, like the ETSI conformance test suites, could also be used to test the simulation model.

Therefore, we integrated an interface module in the INET simulation model which allows communication between the simulated nodes and real nodes connected to the host running the simulation via an IP-based network. This capability proved to be very helpful during the analysis and testing of the SCTP simulation model.

As the formats of packets used in the simulation differ from the ones sent on the wire there have to be two methods to convert the simulated messages to the real packets and vice versa. We called them *Serializer* and *Parser*.

The *Serializer*, reading the data from the simulation and writing them on the wire, also provides an additional value for the dump module. As the dump output can only be traced in fast mode, a mechanism is needed to store the simulation traces. Therefore, the packets are converted by the *Serializer* with additional headers to the

Table 7.1 Configurable parameters according to RFC 4960

Parameter	Meaning	Default Value
assocMaxRetrans	Maximum number of consecutive unacknowledged heartbeats and retransmissions, before the peer is considered to be unreachable	10
hbInterval	Interval between two heartbeats	30 s
maxInitRetrans	Maximum number of retransmissions for an INIT chunk	8
pathMaxRetrans	Maximum number of consecutive unacknowledged heartbeats and retransmissions on a certain path, before the path is set inactive	5
rtoAlpha	Needed to calculate RTO	$\frac{1}{8}$
rtoBeta	Needed to calculate RTO	$\frac{1}{4}$
rtoInitial	Initial retransmission timeout	3 s
rtoMin	Minimum of the retransmission timeout	1 s
rtoMax	Maximum of the retransmission timeout	60 s
validCookieLifetime	Lifespan of the State Cookie	10 s

pcap format and stored in files. Thus they can be analyzed by the Wireshark packet analyzer (see Section 7.5). A user can turn this feature on by providing a name for the trace file. For more details refer to Tüxen et al. (2008a).

7.3.4 Configurable Parameters

In order to be able to test the network under different conditions there are specific parameters that can be easily configured by editing a text file. They are divided into those concerning the transport layer and those specifying the applications.

SCTP parameters

The parameters concerning the transport layer can be configured for each host individually. In Stewart (2007) a number of parameters with their default values are listed, that can be set by the user. They have their equivalents in the following alphabetically ordered list. Most implementations provide the user with additional SCTP kernel parameters. We made the following attributes configurable.

Table 7.2 Additional configurable SCTP parameters

Parameter	Meaning	Default Value
sctpAlgorithmClass	Superclass of the SCTP Algorithm	"SCTPAlg"
arwnd	Advertised receiver window to be announced in the INIT or INIT-ACK chunk	65535
maxBurst	Maximum number of packets that may be sent at once	4
nagleEnabled	Indicates whether the Nagle algorithm is used or not	true
naglePoint	Number of bytes when a packet is considered to be full and can be sent, when the Nagle algorithm is enabled	1468
numGapReports	Number of SACKs that have to report this data chunk to be missing before it is fast retransmitted	4
reactivatePrimaryPath	Indicates whether the original primary path should be activated after it has lost its status and has come up again	false
sackFrequency	Number of chunks to arrive before a SACK is sent	2
sackPeriod	Time after which the sack timer expires and a SACK has to be sent	200 ms
ccModule	Indicates which congestion control model should be used	0 = RFC4960
ssModule	Indicates which stream scheduling model should be used	0 = ROUND_ROBIN
swsLimit	Silly window syndrome avoidance limit. For advertised receiver windows smaller than swsLimit, a window of 1 is announced	3000
sendQueueLimit	Maximum size of the send queue	0

SCTPApp parameters

In addition to those parameters that are needed for every connection like destination *address*, *primaryPath*, number of *outboundStreams* or *message length*, some parameters are provided that allow to change the sending behavior (*delayFirstRead*, *thinkTime* and *echoDelay*). To allow more predictable and longer testing times the parameters *queueSize*, *startTime* and *stopTime* can be configured. As a host can start several applications, so that for instance a server can have associations with numerous clients, the application parameters can be configured independently for every application.

7.3.5   Testing the Simulation

As a simulation is self-contained, the question arises how it can be tested, to evaluate its correctness.

The obvious way is to configure scenarios, test them and see whether the results are plausible.

A more systematic approach is the use of test cases. As we implemented the external interface, it was possible to connect to an external computer. An SCTP testtool and the corresponding ETSI conformance tests are provided in Tüxen (2011). The testtool ran on the external computer and used the simulation as an SUT (system under test). After having passed the tests we were sure that the most important features of the protocol were implemented correctly.

But still the RFCs leave the possibility to interpret some specifications in different ways. Therefore, interoperability events bring developers of various implementations together to test their products against each other. In 2006, we attended the 8th SCTP Interoperability Test in Vancouver, Canada, and in 2007 the 9th SCTP Interoperability Test in Kyoto, Japan. Each time we had the chance to enhance our simulation and make it more robust.

7.4   Configuration Examples

7.4.1   Setting up Routing for a Multihomed Host

One important feature of SCTP is multihoming, which allows the sending of the data belonging to one association over two different paths. In Figure 7.8 a simple network setup is to be seen. INET offers the `FlatNetworkConfigurator`, a module, that automatically assigns IP-addresses from the same subnet to the interfaces. As the addresses on a multihomed host should belong to different subnets, routing files are necessary to set the IP-addresses and the routes for the network devices. In the following the routing behavior of `SCTPClient` will be explained. First the routing file `client.mrt` is defined in `Multihomed.ned` that describes the network.

```
SCTPClient: StandardHost {
    parameters:
```

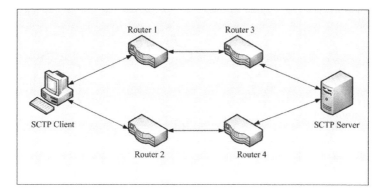

Figure 7.8 Scenario with a multihomed client and server.

```
            IPForward = false;
            routingFile = "client.mrt";
            @display("p=61,142;i=device/laptop");
        gates:
            pppg[2];
    }
```

Then the IP-addresses and the routes are specified in the routing file `client.mrt`:

```
ifconfig:

name: ppp0 inet_addr: 10.1.1.1 Mask: 255.255.255.0 MTU: 1500
Metric: 1 POINTTOPOINT MULTICAST
name: ppp1 inet_addr: 10.2.1.1 Mask: 255.255.255.0 MTU: 1500
Metric: 1 POINTTOPOINT MULTICAST

ifconfigend.

route:

10.1.0.0   10.1.1.254   255.255.0.0   G   0   ppp0
10.2.0.0   10.2.1.254   255.255.0.0   G   0   ppp1

routeend.
```

These two steps have to be completed for each device. As a consequence, INET will read the routing files and take them as basis for the routing decisions.

7.4.2   Connecting an External Interface to the Real World

At present the external interface is integrated in the module `ExtRouter` and the `StandardHost`. Thus it is possible to connect just one host to the outside world

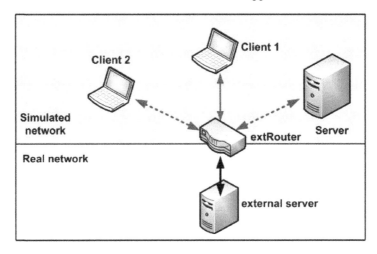

Figure 7.9 Scenario with an external router.

or a complete simulated network. Figure 7.9 shows a scenario, where Client2 is connected to the Server, and Client1 is configured to send data to the external server. The extRouter belongs to both 'worlds', as it accepts the pure simulation data from Client2 and Server and the messages being exchanged between Client1 and the external server. The routing file for extRouter looks as follows:

```
ifconfig:

name: ppp0  inet_addr: 10.1.1.254 Mask: 255.255.255.0 MTU: 1500
Metric: 1 POINTTOPOINT MULTICAST
name: ppp1  inet_addr: 10.1.2.254 Mask: 255.255.255.0 MTU: 1500
Metric: 1 POINTTOPOINT MULTICAST
name: ppp2  inet_addr: 10.1.3.254 Mask: 255.255.255.0 MTU: 1500
Metric: 1 POINTTOPOINT MULTICAST

# real ethernet card (going on the wire)
name: ext0 inet_addr: 10.1.4.254 Mask: 255.255.255.0 MTU: 1500
Metric: 1 POINTTOPOINT MULTICAST

ifconfigend.

route:
10.1.1.1   10.1.1.1    255.255.255.255 H   0    ppp0 #client1
10.1.2.1   10.1.2.1    255.255.255.255 H   0    ppp1 #client2
10.1.3.1   10.1.3.1    255.255.255.255 H   0    ppp2 #server

0.0.0.0 *     0.0.0.0  G   0    ext0
```

`routeend.`

The next step is to set the filters in the configuration file, e.g., `omnetpp.ini`.

```
scheduler-class = "cSocketRTScheduler"
**.ext[0].filterString = "ip and (dst net 10.1.0.0/16)"
**.ext[0].device = "eth0"
```

The filter catches all IP traffic that is bound for either the simulated hosts, which is the normal case for an SCTP association, or `extRouter` itself, in case the ping command is to be used between the router and the external server. "eth0" has to be substituted by the name of the computer's real network adapter. The last step is to add a route on the `external server` to the simulation via the local computer, where the simulation is running. Assuming the IP-address of the local computer to be '192.168.1.10', the command to add the route could be:

```
route add -net 10.1.0.0 netmask 255.255.0.0 gw 192.168.1.10
```

Now everything is ready to set up an association between the simulation and `external server`.

7.5 Using the Wireshark Packet Analyzer for the Graphical Analysis of the Data Transfer

Large trace files can become unmanageable fast, so that analyzing them to find faulty behavior is almost impossible. Therefore, a graphical tool to visualize the data transfer is very helpful. Besides the visualization of the data flow, statistical data, like the number of chunks can lead to a better understanding, too.

Wireshark is an open source packet analyzer available from Combs (2011). The traces which can be visualized with Wireshark not only stem from the INET simulation, but all SCTP trace files captured with Wireshark, dumpcap, tcpdump, or snoop can be graphically analyzed.

After having opened the capture file, there are two possibilities to start the analysis. To get information about all associations traced the menu entry *Telephony - SCTP* leads to the choices

- Analyze this Association

- Chunk Counter...

- Show All Associations...

Via the context menu of an SCTP packet the actual association can be directly analyzed (*SCTP - Analyze this Association*) or a filter can be prepared (*SCTP - Prepare Filter for this Association*), that can be applied to the trace to see only association related frames.

In the following, after a short explanation, how packets can be assigned to associations, the different ways to analyze the traffic will be discussed.

7.5.1   Assigning Packets to Associations

To be able to analyze the data transfer, the packets have to be mapped to associations. The usual way to do this is to match source address and source port with destination address and destination port. As a tool like Wireshark is also used to develop protocols and detect bugs, the data sent often stem from test cases, where the same port numbers are chosen for each association. The top window of Figure 7.10

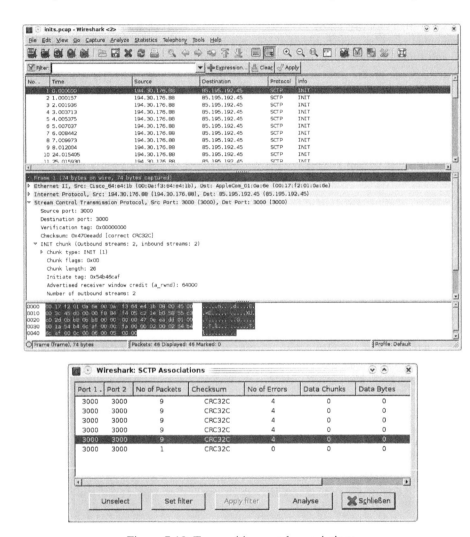

Figure 7.10 Trace with several associations.

shows a trace, where only INIT-chunks were transmitted belonging to six different associations (see bottom window of Figure 7.10). Source and destination ports are

Table 7.3 Assigning addresses and verification tags to associations

chunk type	source address	local v_tag	destination address	remote v_tag
INIT	10.0.0.1	0	10.0.0.2	12345
INIT-ACK	10.0.0.1	**56789**	10.0.0.2	12345

the same in all packets. Therefore, it is not possible to make a statement concerning the association, to which they belong.

An SCTP specific variable is the verification tag, that is unique for either side of an association. Thus the combination of local verification tag, source address to remote verification tag and destination address is a good way to identify an association.

The SCTP handshake is the basis for the association. In the INIT-chunk the initiation tag informs the receiver, which verification tag it has to send. In Table 7.3 an entry reflects the necessary combination of verification tags and addresses to identify an association. With the arrival of an INIT-chunk already four fields can be filled. The remote verification tag is equal to the initiate tag of the INIT-chunk. Only the local verification tag is not known yet. This information is provided by the initiate tag of the INIT-ACK-chunk. All other chunks can be easily assigned, as all necessary values are present.

The situation is more difficult, when only part of a trace is available or the handshake is not complete like in the example in Figure 7.10. Assuming two INIT-chunks arriving from opposite directions. The first one provides source address, local v_tag and destination address. From the second one the addresses fit, but we do not know whether the verification tag is the right one. In this case it can be helpful to also compare the port numbers to exclude the possibility, that the chunks belong to different associations.

While reading the trace file, all information necessary for later analysis, like the chunks, maximum and minimum values for TSNs and time, number of bytes, chunks, etc., is stored in a structure. The different associations are organized in a list.

7.5.2 Statistics

There are two different chunk statistics implemented for SCTP in Wireshark. One is reached via the main menu and the other one from the SCTP Analyze Association window. The first way leads to the top window of Figure 7.11. Without putting in a filter string, all associations are analyzed as seen in the second window of Figure 7.11. The way, packets are assigned to associations differs from the one described in Section 7.5.1, so that in the case when port numbers do not change from one association to another, the associations cannot be distinguished.

There are four different ways to fill in the filter string text field of the small window of the chunk counter:

- typed in manually,

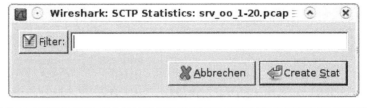

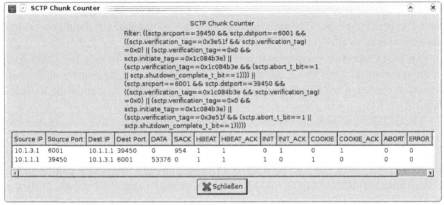

Figure 7.11 SCTP chunk counter.

- with predefined filter functions chosen from the Display Filter Dialog reached via the Filter button,

- via the function *SCTP - Prepare Filter for this Association* in the context menu of an SCTP packet,

- via one of the *Set Filter* buttons.

The result for an example can be seen in the bottom window of Figure 7.11. To reach the second kind of chunks statistics either an association of the entries in the SCTP Associations window can be chosen (see Figure 7.10), *Analyze this Association* can be called from the main menu or the context menu of an SCTP packet. The first dialog window that appears is the top one of Figure 7.12. It lists some statistical data, that are relevant for the association like errors and the number of DATA-chunks and bytes in either direction.

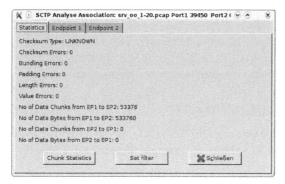

ChunkType	Association	Endpoint 1	Endpoint 2
DATA	53376	53376	0
INIT	1	1	0
INIT_ACK	1	0	1
SACK	954	0	954
HEARTBEAT	2	1	1
HEARTBEAT_ACK	2	1	1
ABORT	0	0	0
SHUTDOWN	0	0	0
SHUTDOWN_ACK	0	0	0
SCTP_ERROR	0	0	0
COOKIE_ECHO	1	1	0
COOKIE_ACK	1	0	1
ECNE	0	0	0
CWR	0	0	0
SHUT_COMPLETE	0	0	0
AUTH	0	0	0
NR_SACK	0	0	0
Others	0	0	0

Figure 7.12 Statistics of the chunk types.

For more information, the Chunk Statistics button can be clicked to open the bottom window of Figure 7.12. For each important chunk type its number per endpoint and for the complete association is listed.

7.5.3   Graphical Representation of the Data Transfer

The graphical analysis of the data transfer is performed on a per endpoint basis. Therefore, the user has to select an endpoint via the tabs of the top window of Figure 7.12 and choose the graphical presentation he prefers.

Analyzing TSNs and SACK-chunks

To be able to see the course of the TSNs it is very helpful to analyze congestion control issues. Figure 7.13 shows an example. Besides the TSNs, the cumulative TSN

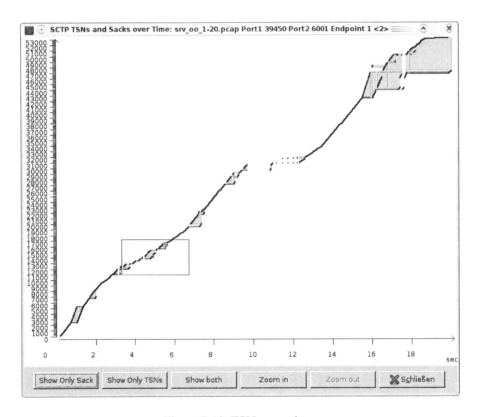

Figure 7.13 TSNs over time.

Acks and the TSNs that have been announced in the Gap Ack Blocks are visible, too. To see only the TSNs, taken from the DATA-chunks or only those from the SACK-chunks, the *Show Only TSNs* or *Show Only Sacks* button can be clicked. To zoom into the graph a rectangle must be drawn around the area to be magnified (see Figure 7.13). With a click on the *Zoom in* button or into the rectangle, a detailed figure appears, like the one in Figure 7.14. Now the gaps in the course of the TSNs and

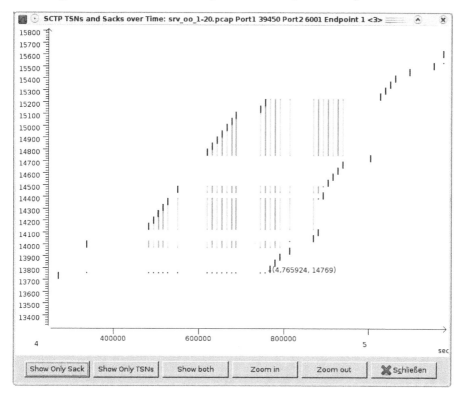

Figure 7.14 Clipping of TSNs over time.

retransmissions can easily be pointed out. In the case of multihomed hosts, timer-based retransmissions are not visible, because they are sent on the second path and therefore traced on another interface.

For convenience, the user can pick out one TSN and see its coordinates by clicking on it. A double-click selects the corresponding frame in the main window.

Analyzing the Advertised Receiver Window and Transmitted Bytes

Another important feature of SCTP besides congestion control is flow control. To analyze flow control scenarios the size of the advertised receiver window is very important. Choosing the button *Graph Bytes* leads to a window like the one in Figure 7.15. The course of the DATA-chunks can be observed again, but this time the y-axis shows the accumulated number of bytes instead of the TSNs. The dark gray lines represent the arwnd. As this is the representation corresponding to the one in Figure 7.13, the relation between the rise and fall of the arwnd and the data stored at the receiver, is well to be seen. The large Gap Ack Blocks in Figure 7.13, starting at about 15 secs, indicate the number of acknowledged TSNs, that can not be delivered

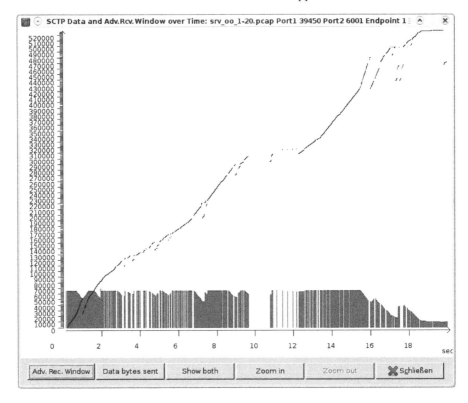

Figure 7.15 Advertised receiver window and transmitted bytes.

to the application, because in-order delivery is not possible due to missing TSNs. Therefore, the data has to be cached in the receive queues which reduces the arwnd, as seen in Figure 7.15. The intermediate rise results from the reception of missing TSNs, that led to an increase of the cumulative TSN ack parameter.

7.6   Conclusion

This chapter discussed the SCTP simulation model that is integrated in the INET framework. It can be used to analyze network scenarios with nodes using SCTP as the transport protocol. Multihoming as an integral part of SCTP can be tested and evaluated. A special feature of our implementation is the external interface that can be used to connect the simulation to real computers and even networks. Thus, new features can be tested in conjunction with real network equipment. The trace files that are gathered thanks to an additional dump module can be analyzed with the Wireshark network packet analyzer and their flows can be visualized.

Bibliography

Combs G 2011 Wireshark protocol analyzer. Available at: http://www.wireshark.org.

Nagle J 1984 Congestion Control in IP/TCP Internetworks. *RFC 896.*

Rüngeler I 2009 *SCTP - Evaluating, Improving and Extending the Protocol for Broader Deployment* PhD thesis University of Duisburg-Essen.

Stewart R 2007 Stream Control Transmission Protocol. *RFC 4960.*

Stewart R, Arias-Rodriguez I, Poon K, Caro A and Tüxen M 2006 Stream Control Transmission Protocol (SCTP) specification errata and issues. *RFC 4460.*

Stewart R, Xie Q, Tüxen M, Maruyama S and Kozuka M 2007 RFC5061: Stream Control Transmission Protocol (SCTP) Dynamic Address Reconfiguration. *Internet RFCs.*

Tüxen M 2011 SCTP Testtool (stt). Retrieved from: http://sctp.fh-muenster.de/sctp-testtool.html.

Tüxen M, Rüngeler I and Rathgeb E 2008a Interface connecting the INET simulation framework with the real world *Simutools '08: Proceedings of the 1st international conference on simulation tools and techniques for communications, networks and systems & workshops.*

Tüxen M, Rüngeler I, Stewart R and Rathgeb E 2008b Network Address Translation for the Stream Control Transmission Protocol. *IEEE Network* **22**(5), 26–32.

Tüxen M, Stewart R, Lei P and Rescorla E 2007 RFC4895: Authenticated Chunks for the Stream Control Transmission Protocol (SCTP). *Internet RFCs.*

Varga A 2010 The OMNeT++ Discrete Simulation Environment. Retrieved from: http://www.omnetpp.org.

Varga A 2011 INET Framework. Retrieved from: http://inet.omnetpp.org.

Varga A and Hornig R 2008. An overview of the OMNeT++ simulation environment. *Simutools '08: Proceedings of the 1st international conference on simulation tools and techniques for communications, networks and systems & workshops.*

Wilson M 2011 Eclipse Retrieved from: http://www.eclipse.org.

8

SCTP Application Interface

Lode Coene

This chapter describes what sort of data exchanges between the peers is possible, dependencies between data within an association, and how to do secure data exchange over your association, and demonstrates the simple versus rich interface models of SCTP between the application and the transport layer.

8.1   Interface between the Transport Protocol and Application

The interface between an application and a transport can be closely tied to the operating system running on the peer, but most interfaces are generic across multiple operating systems (Unix, Windows, etc.).

This allows us to describe an generic interface used by the application and provided by the transport. Some parts of the interface are the same as already existing UDP and TCP provided interfaces. It even allows SCTP to mimic the interface as long as specific SCTP features are not used.

One of the most important choices for an application is to choose how to exchange data between the peers. Basically there are two different ways of communication used nowadays.

- Message based: delimited messages are exchanged between the 2 peers. The information contained in the message is self-described and has a known length before it is transmitted towards the remote end. The transport knows the exact start and stop of the information in sending and receiving direction.

- Stream based: a stream of bytes is exchanged between the 2 peers. The information is not always self-described and does not always have a known length before it is transmitted towards the remote end. Information may be framed by the application within the stream but the transport has absolutely no idea where the start and/or the end of the information is in sending or receiving direction.

Message-based communication is used by applications which know before they send out the information how long the information is. Stream-based communication is, in general, used by applications which do NOT know the length of the information in advance. The application will know the length at a certain point but not before (at that moment some of the data may have already been transmitted).

Another way of looking at information exchange is the handling of time within the transmission. If asynchronous transmission is employed, then the data are transmitted between the peers with no time restrictions placed on it. The data may therefore take a long time in transit between the peers.

If, on the other hand, synchronous transmission is employed then the data are transmitted between the peers with a time restriction placed on them. It has to reach the remote peer at that time or must be discarded.

SCTP will, in most cases, be employed for asynchronous communication, but a subset of application can use the partial reliability mode of SCTP to have synchronous communication with the remote peer.

The application can decide to employ one or more streams within an SCTP association.

If one stream is used, then all messages or streams will be sequentially sent and received as in a First in First Out (FIFO) order. If message loss on the lower layers is observed, this may delay subsequent messages. If no dependencies exist between 2 messages, then the 2 messages can be sent on different streams of the association. The delay of the first message will not have an impact anymore on the delivery of the second message.

When the unordered delivery of SCTP is selected, then messages may even overtake each other within a stream of an SCTP association when the first messages are delayed in the stream.

The impact of application layer framing

Information can be framed by the application prior to sending it to the transport. This can be done in message-based or stream-based transport mode. The framing of the message is easier when using message-based communication as the information must be contained in the message itself. This allows for some synchronization help by the transport to the application. If a message is not complete (as required by the application protocol) when receiving, then the next message can still be received and processed if it conforms to the application protocol.

This is not the case in stream-based communication where the synchronization is completely dependant on the application with no help from the transport layer. If the application loses the synchronization (it does not know where the next frame is in the stream), then the application is forced to terminate the SCTP association.

8.2 General SCTP API Functions

The SCTP socket API is described in general in SCTPref while the more practical use of the API is decribed in UnixNetProg.

SCTP has 2 ways of working with sockets:

- One-to-one style: each association gets its own socket descriptor

- One-to-many style: one socket descriptor handles many associations

The one-to-many style allows one to save on processing of socket descriptors. The specific association is no longer selected via the socket but via the association identifier, which is carried along any request or indication to/from the socket.

The one-to-one style corresponds to the more traditional way of handling UDP and TCP transport data exchange and allows a TCP or UDP connection to be ported to SCTP without too much hassle.

The following functions are used for setting SCTP data exchange with a remote peer.

Socket

The socket function specifies which sort of protocol will be used between the peers.
Res = socket (family, type, protocol)

1. The family specifies the sort of network you want to use IPv4 or IPv6. More exotic choices are Unix domain protocol and routing protocols.

2. The type specifies the sort of data interface the application is going to use.

   (a) SOCK_STREAM: stream-based data exchange

   (b) SOCK_SEQPACKET: datagram exchange

A socket with SOCK_SEQPACKET indicates that the communication will use sequenced datagram delivery only but the send function will allow the application to use unsequenced datagram delivery. This allows an application to use a single association for sequenced and unsequenced message delivery.

1. The protocol indicates which protocol is going to be used. SCTP associations always use IPPROTO_SCTP value.

The result of the SOCKET call is a socket descriptor. The socket descriptor fully (one-to-one style) or partially (one-to-many style) identifies the association on which the data can be exchanged.

Bind

The bind function indicates to which network interfaces and portnumber an application is bound to the socket descriptor. One or more interfaces with its corresponding IP address and portnumber can be added to or removed from a single socket. TCP or UDP sockets bind to a single address or to all addresses present. SCTP allows to selectively choose to which address/interface to add to the socket or remove an interface from the socket.

Res = Bind(sockfd, sockaddr, addresslen)

Res = Bindx(sockfd, sockaddrlist, addresscnt, flags)

The bind function is used in the both one-to-one and one-to-many style socket.

Listen

The listen function indicates that a certain socket will be listening for incoming associations on the address and port specified in the bind function. This allows computers to act as servers for incoming associations. The counterpart to the server (= the client) will be using the send or connect function to initiate the association with the remote server.

Res = Listen(sockfd, backlog)

If an association arrives on a SOCK_STREAM socket it must be accepted via the function accept.

If an association arrives on a SOCK_SEQPACKET socket the message can be received via the recvmsg function.

The listen function is used in the both one-to-one and one-to-many style socket.

Accept

The accept function accepts incoming SCTP SOCK_STREAM-based associations on a listening SCTP socket.

Res = accept(sockfd, sockaddr, addresslen)

The accept function is only used in the one-to-one style socket.

8.3   General SCTP Handling Functions

The following functions show a test application written on top of transport layer sockets. This allows us to show the different ways in which to use SCTP in different environments. Some of functionality will not be described but will be included in the source code attached to the book.

The main function contains the initialization and the infinite loop for polling the sockets.

The eventloop function is a select statement on any socket looking for events from the sockets. When an IP packet is received, the socket on which the message was received will become ready. Then code must be called for receiving and handling any events or data from that particular socket. The events must be examined for the different event types that SCTP can generate. Some events are internal to SCTP while other events indicate that data are received from the remote peer.

Depending on the mode selected on the socket, only data may be received or data and SCTP events can be received by the application.

```
{
  int       res;

  cout << "Welcome to Testtool\n";
  // initialize testtool(and SCTP)
  res = test_initialisation();
  // register testool to the sualibrary and its users
  res = testtool_unix_init();

  /* analyse the commandfile(if present) */
  res = test_AnalyseCfile(argc, argv[1], argv[2]);

  // go into infinite loop( = handle all events)
  while ((res = test_eventLoop()) >= 0);

  // close the program
  exit (0);

}
```

Figure 8.1 Test main program.

After checking if enough time has elapsed, check which sort of filedescriptors is used. A filedescriptor can become ready for the following situations:

- reading from the filedescriptor

- writing to the filedescriptor

- an exception generated by the filedescriptor

```
{
  int                  result ,n,i;
  fd_set               read_fd_set , write_fd_set , exception_fd_set ;
  struct timeval       timeout;
  struct timeval       *to;
  int                  msecs = -1;

  /* initialise structures for select */
  n = 0;
  FD_ZERO(&read_fd_set );
  FD_ZERO(&write_fd_set );
  FD_ZERO(&exception_fd_set );

  if (msecs < 0)
    { /* select till something comes up */
      to = NULL;
    }
  else if (msecs == 0)
    { /* poll the fd */
      test_dispatch_timer ();
      to = &timeout;
      timeout.tv_sec = 0;
      timeout.tv_usec = 0;
    }
  else
    { /* set the timer */
      to = &timeout;
      timeout.tv_sec = msecs / 1000;
      timeout.tv_usec = (msecs % 1000) * 1000;
    }
```

Figure 8.2 test_eventloop function initialization.

The select statement will wait until one of the filedescriptors becomes ready. Then the set of descriptors is checked to see which filedescriptor became ready. It could be that more than one descriptor became ready.

```
  for (i = 0; i < num_of_fds; i++)
    {
      if (poll_fds[i].fd < 0)
        continue;
      n = max(n, poll_fds[i].fd);
      /* filedescriptor waiting / looking for reading ? */
      if (poll_fds[i].events & (POLLIN | POLLPRI))
        FD_SET(poll_fds[i].fd, &read_fd_set );
      /* filedescriptor waiting / looking for writing ? */
      if (poll_fds[i].events & (POLLOUT))
        FD_SET(poll_fds[i].fd, &write_fd_set );
      /* filedescriptor waiting / looking for exceptions ? */
      if (poll_fds[i].events & (POLLIN | POLLOUT))
```

```
        FD_SET( poll_fds[i].fd, &exception_fd_set );
```
Figure 8.3 test_eventloop function filedescriptor read/write/exception mechanism.

```
result = select( n + 1,
                 &read_fd_set,
                 &write_fd_set,
                 &exception_fd_set,
                 to
               );

if (result > 0)
  {
    /* do the handling of every filedescriptor that came */
    /* up during this select */
    for (i = 0; i < num_of_fds; i++)
      {
        /* set recorded event to 0 */
        poll_fds[i].revents = 0;
        if ((poll_fds[i].events & POLLIN) &&
            (FD_ISSET(poll_fds[i].fd, &read_fd_set)))
          {
            /* cout << "select did catch something\n ";*/
            poll_fds[i].revents |= POLLIN;
          }

        if ((poll_fds[i].events & POLLOUT) &&
            (FD_ISSET(poll_fds[i].fd, &write_fd_set)))
          {
            /*cout << "select did catch something\n ";*/
            poll_fds[i].revents |= POLLOUT;
          }

        if ((poll_fds[i].events & (POLLIN | POLLOUT)) &&
            (FD_ISSET(poll_fds[i].fd, &exception_fd_set)))
          {
            poll_fds[i].revents |= POLLERR;
          }

      }
```

Figure 8.4 test_eventloop function select.

```
/* do the actual handling (...finally ...) */
   for (i = 0; i < num_of_fds; i++)
     {
       /* was their a event on this filedescriptor? */
       if (poll_fds[i].revents != 0)
         {
           if ((poll_fds[i].revents & POLLPRI) ||
               (poll_fds[i].revents & POLLIN) ||
               (poll_fds[i].revents & POLLOUT))
```

```
{
    /* activity on filedescriptor !!! */
    /* user ? */
    if (( event_callbacks [ i ]−>event_tcb_type == EVENT_TYPE_USER) &&
        ( event_callbacks [ i ]−>action != NULL))
    {
        /* cout << "user callback called for normal processing\n "; */
        (( test_userCallback )*
          ( event_callbacks [ i ]−>action )) ( poll_fds [ i ]. fd ,
                                                poll_fds [ i ]. revents ,
                                                &poll_fds [ i ]. events ,
                                                event_callbacks [ i ]−>userData
                                              );
    }
```

Figure 8.5 test_eventloop function user event on filedescriptor.

```
/* SCTP ? */
                    if (( event_callbacks [ i ]−>event_tcb_type == EVENT_TYPE_SCTP ) &&
                        ( event_callbacks [ i ]−>action != NULL))
                    {
                        /* cout << "test callback called for  SCTP msg processing\n "; */
                        (( sua_userCallback )*
                          ( event_callbacks [ i ]−>action )) ( poll_fds [ i ]. fd ,
                                                                poll_fds [ i ]. revents ,
                                                                &poll_fds [ i ]. events ,
                                                                event_callbacks [ i ]−>userData
                                                              );
                    }
                }
```

Figure 8.6 test_eventloop function SCTP event on filedescriptor.

```
    if (( poll_fds [ i ]. revents & POLLERR ))
                    {
                        /* error condition on filedescriptor */
                        if (( event_callbacks [ i ]−>event_tcb_type == EVENT_TYPE_USER) &&
                            ( event_callbacks [ i ]−>action != NULL))
{
                            (( sua_userCallback )*
                              ( event_callbacks [ i ]−>action )) ( poll_fds [ i ]. fd ,
                                                                    poll_fds [ i ]. revents ,
                                                                    &poll_fds [ i ]. events ,
                                                                    event_callbacks [ i ]−>userData
                                                                  );
                        }
                    }
                }

            /* else continue to next filedescriptor */
            /* reset received events for next run */
            poll_fds [ i ]. revents =0;

        }
    }
    else  if ( result == 0)
    {
        /* do the timer handling: timer expired on this select */
        test_dispatch_timer ();
    }
    return result;
}
```

Figure 8.7 test_eventloop function exception event and next filedescriptor run.

The eventloop checks for error conditions that may arise on socket descriptors. Then it will process the next ready filedescriptor. The loop will continue until a timeout is generated by the test_dispatch timer. This allows for nonfiledescriptor-based input/output to occur.

8.4 Message-Based Communication

Simple datagram association

UDP applications can be easily converted to the simple model of an SCTP message-based association.

The communication between the peers is set up using the SCTP protocol, but to the application, it looks as if it is using a UDP socket. The application uses only the send and receive function of the API. Any transactional housekeeping needed for keeping some state across subsequent messages to the peer is done by the application.

If subsequent messages are sent to the same remote peer and the association has not been torn down yet, then the messages will be transmitted across the already active association. The period that an association remains open to the remote peer can be configured.

Another possibility for optimization is when the application message is sent on the third message of the SCTP association setup sequence (actually it is the second message sent from the initiator peer to the remote peer). Normally application data are only sent on the fifth message of the association, just after the four-way handshake of the SCTP setup has been completed.

```
bool   test_create_SCTP_socket_sipmle( int  i_host_idx )
{
  struct sctp_event_subscribe  events;
  unsigned int  i;
  int     res;
  /* open the SCTP socket and get the filedescriptor */
  hostTable[ i_host_idx ]. sctpInstance = socket( AF_INET,
                                    SOCK_SEQPACKET,
                                    IPPROTO_SCTP
                                  );

  if (hostTable[ i_host_idx ]. sctpInstance < 0)
    {
      return false;
    }

  /* register the test callback functions for the SCTP filedescriptor */
  /* so if a msg is received by SCTP, it can be handled by the */
  /* test callback function */
  res = test_register_fd( hostTable[ i_host_idx ]. sctpInstance ,
                          EVENT_TYPE_SCTP,
                          POLLIN | POLLPRI,
                          (void *) &test_handle_sctp_msg ,
                          NULL
                        );
  if (res < 0)
    {
      return false;
    }

  if (hostTable[ i_host_idx ]. localIpAddr[0]. ss . ss_family == AF_INET) {
```

```
                bind ( hostTable [ i_host_idx ]. sctpInstance , ( const struct  sockaddr *)
                          & hostTable [ i_host_idx ]. localIpAddr [0] ,
                                  sizeof ( struct  sockaddr_in ));
    }
else {
                bind ( hostTable [ i_host_idx ]. sctpInstance , ( const struct  sockaddr *)
                          & hostTable [ i_host_idx ]. localIpAddr [0] ,
                                  sizeof ( struct  sockaddr_in6 ));
    }
}
```

Figure 8.8 Create simple SCTP UDP like socket.

After creating the socket, messages can be sent or received between the peers as shown in the following figures. The association will be a single-homed association. The remote address must be specified in the send as no connect has been issued by the client for setting up the association.

```
Bool   test_send_msg_UDP ( int              assoc_id ,
                           data_str     *i_msg_name_ptr ,
                           int          outstream )
{
    int              msg_index ;
    int              result ;
    int              assoc_index ;
    std :: string    temp ;
    std :: string    msg_name ;
    char             temp_char [1000] ;
    char             temp1 [5] ;
    int              total_length = 0 ;
    char             *disp_msg_ptr ;

    /* copy msg into a temp var */
        char * databuf = new char [ i_msg_name_ptr ->length + 1 )] ;
        memcpy ( ( char *) databuf ,
                & i_msg_name_ptr ->contents ,
                i_msg_name_ptr ->length
                ) ;

        // send message
        if ( assocLinkTable [ i_assoc_idx ]. destIpAddr [0]. ss . ss_family == AF_INET) {
                result = sendto ( assocLinkTable [ assoc_index ]. sctpInstance ,
                        ( void *) databuf ,
                        i_msg_name_ptr ->length
                    MSG_WAITALL,
                            ( struct  sockaddr *)
                                &assocLinkTable [ i_assoc_idx ]. destIpAddr [0] ,
                        sizeof ( struct  sockaddr_in )) ;
        }
        else {
                result = sendto ( assocLinkTable [ assoc_index ]. sctpInstance ,
                        ( void *) databuf ,
                        i_msg_name_ptr ->length
                    MSG_WAITALL,
                            ( struct  sockaddr *)
                                &assocLinkTable [ i_assoc_idx ]. destIpAddr [0] ,
                        sizeof ( struct  sockaddr_in6 )) ;
        }

    /* delete the temporary buffer */
    delete databuf ;

    return true ;
}
```

Figure 8.9 Simple SCTP UDP-like send data.

```
void test_handle_sctp_msg_UDP ( int   fd ,   short int   revents ,   short int   *gotEvents ,
void   *dummy )
{
  int                        msg_flags , msg_len ,  addr_length ;
  int                        nr_destinations ;
  socklen_t                  peer_addr_len ;
  char                       readbuf [4096] ;

  memset ( readbuf ,   0 , sizeof ( readbuf )) ;

  msg_len  =  recvfrom ( assocLinkTable [ assoc_index ] . sctpInstance ,
                        readbuf ,
                        sizeof ( readbuf ) ,
                       &msg_flags ,
                           ( struct  sockaddr *)
                               &assocLinkTable [ i_assoc_idx ] . destIpAddr [0] ,
                        addr_length
               ) ;


      test_DistributeMsg_simple (              Readbuf ,
                                                         msglen
                                      ) ;


}
```

Figure 8.10 Simple SCTP UDP-like data receive.

Normal 2 way datagram association

TCP applications which employ application layer framing can convert to the normal model of an SCTP message-based association.

This mode is actually a combination of TCP and application layer framing, but as SCTP does the framing anyway, the application code looks a bit less complex or does not need to employ framing.

The application is interested in setting up/connecting to the remote application, send and receiving data and, finally, closing the association with its remote peer.

The application uses the connect function to initiate the four-way handshake with the remote peer.

Once the connection is set up, messages can be sent using the normal send function available in the socket API.

```
bool   test_create_SCTP_socket_simple ( int  i_host_idx )
{
  struct  sctp_event_subscribe   events ;
  unsigned  int  i ;
  int    res ;
  /* open the SCTP socket and get the filedescriptor */
  hostTable [ i_host_idx ] . sctpInstance  =  socket ( AF_INET ,
                                                SOCK_SEQPACKET ,
                                                IPPROTO_SCTP
                                                ) ;

  if  ( hostTable [ i_host_idx ] . sctpInstance  <  0 )
    {
      return  false ;
    }

  /* register the test callback functions for the SCTP filedescriptor */
```

```
/* so if a msg is received by SCTP, it can be handled by the */
/* test callback function */
res = test_register_fd ( hostTable[i_host_idx].sctpInstance,
                         EVENT_TYPE_SCTP,
                         POLLIN|POLLPRI,
                         (void *) &test_handle_sctp_msg,
                         NULL
                       );
if (res < 0)
  {
    return false;
  }

if (hostTable[i_host_idx].localIpAddr[0].ss.ss_family == AF_INET) {
    bind(fd, (const struct sockaddr *) & hostTable[i_host_idx].localIpAddr[0],
        sizeof(struct sockaddr_in ));
}
else {
    bind(fd, (const struct sockaddr *) & hostTable[i_host_idx].localIpAddr[0],
        sizeof(struct sockaddr_in6 ));
}
}
```

Figure 8.11 Create simple SCTP socket.

Connect to or accepting an SCTP association is shown in the following figure.

```
bool connect_accept_assoc_simple (int i_assoc_idx)
{
  //association ID not avalaible
  assocLinkTable[i_assoc_idx].assocSctpId = -1;

  if (assocLinkTable[i_assoc_idx].init)
{

    if (assocLinkTable[i_assoc_idx].destIpAddr[0].ss.ss_family == AF_INET) {
        connect(fd, (struct sockaddr *)
                & assocLinkTable[i_assoc_idx].destIpAddr[0],
                    sizeof(struct sockaddr_in ));
    } else {
        connect(fd, (struct sockaddr *)
                & assocLinkTable[i_assoc_idx].destIpAddr[0],
                    sizeof(struct sockaddr_in6 ));
    }
    assocLinkTable[i_assoc_idx].state = ASSOC_ACTIVE;
  }
  else
    {
      assocLinkTable[i_assoc_idx].state = ASSOC_ACTIVE;
      /* listen for any possible incoming associations */
      listen( assocLinkTable[i_assoc_idx].sctpInstance,
          LISTENQ
            );
    }

  return true;
}
```

Figure 8.12 Connect or accept simple SCTP association.

Sending and receiving data are very similar to TCP socket handling as shown in the following figures.

```
Bool  test_send_msg_simple (int           assoc_id,
                        data_str      *i_msg_name_ptr,
```

```
                         int                outstream )
{
  int              msg_index ;
  int              result ;
  int              assoc_index ;
  std :: string    temp ;
  std :: string    msg_name ;
  char             temp_char [1000];
  char             temp1 [5];
  int              total_length = 0;
  char             *disp_msg_ptr ;

  /* copy msg into a temp var */
      char* databuf = new char[i_msg_name_ptr ->length + 1)];
      memcpy ( (char *) databuf ,
              & i_msg_name_ptr ->contents ,
              i_msg_name_ptr ->length
              );

      // send message
      result = send( assocLinkTable [assoc_index]. sctpInstance ,
              (void *) databuf ,
              i_msg_name_ptr ->length
              MSG_WAITALL
              );

      /* delete the temporary buffer */
      delete databuf ;

  return true ;
}
```

Figure 8.13 Simple SCTP send data.

Messages are received via the select function and received by the receive function. As this is datagram mode, the received message length is the actual length of the message as framed and sent by the initiator of the application data exchange.

```
void test_handle_sctp_msg_simple ( int    fd ,  short int  revents ,
                 short int  *gotEvents ,           void    *dummy )
{
  int                        msg_flags , msg_len ;
  int                        nr_destinations ;
  socklen_t                  peer_addr_len ;
  char                       readbuf [4096];

  memset ( readbuf , 0, sizeof ( readbuf ));

  msg_len = recv ( fd ,
                 readbuf ,
```

```
            sizeof(readbuf),
            &msg_flags
        );

    test_DistributeMsg_simple(    Readbuf,
                                                   msglen
                    );


}
```

Figure 8.14 Simple SCTP data receive.

Both peers can close the association using the close function.

When connecting or closing an association towards a remote peer, no indication for a successful or a failed association is delivered to the application to keep the API similar with standard TCP socket programming.

Advanced 2 way datagram association

This employs all of the transport layer indications coming up from the SCTP transport layer.

This mode uses all of the basic features of SCTP and allows one to use the advanced features of SCTP also. A number of extra indications is exchanged between the association and the application. This allows more control of the application for the use of the underlying transport association.

As in any transport, a socket must be created as shown in the following figure:

```
bool    test_create_SCTP_sockett(int i_host_idx)
{
    struct sctp_event_subscribe    events;
    unsigned int i;
    int    res;
    /* open the SCTP socket and get the filedescriptor */
    hostTable[i_host_idx].sctpInstance = socket( AF_INET,
                                                 SOCK_SEQPACKET,
                                                 IPPROTO_SCTP
                                                 );

    if (hostTable[i_host_idx].sctpInstance < 0)
        {
            return false;
        }

    /* set the sort of SCTP notifications that test */
    /* application is interested in */
    bzero(&events, sizeof(events));
    events.sctp_data_io_event = 1;
    events.sctp_association_event = 1;
```

```
events.sctp_shutdown_event = 1;
res = setsockopt( hostTable[i_host_idx].sctpInstance ,
                  IPPROTO_SCTP,
                  SCTP_EVENTS,
                  &events ,
                  sizeof(events)
                );

/* register the test callback functions for the SCTP  */
/* filedescriptor so if a msg is received by SCTP, it */
/* can be handled by the test callback function */
res = test_register_fd( hostTable[i_host_idx].sctpInstance ,
                        EVENT_TYPE_SCTP,
                        POLLIN|POLLPRI,
                        (void *) &test_handle_sctp_msg ,
                        NULL
                      );
if (res < 0)
  {
    return false;
  }
```

Figure 8.15 Create an SCTP socket and register callback function.

```
for (i=0; i< hostTable[i_host_idx].numLocalIpAddrs; i++)
  {
    /* bind each address to the socket */
    res = sctp_bindx( hostTable[i_host_idx].sctpInstance ,
                      (struct sockaddr *) &hostTable[i_host_idx].localIpAddr[i],
                      1,
                      SCTP_BINDX_ADD_ADDR
                    );
    if ( res != 0)
      {
        // not success
        return false;
      }
  }
}
```

Figure 8.16 Bind the local IP addresses to the created SCTP socket.

Connecting to an SCTP remote or listen for incoming SCTP associations is shown in the following figure.

```
bool connect_accept_assoc (int i_assoc_idx)
{
  //association ID not avalaible
  assocLinkTable[i_assoc_idx].assocSctpId = −1;

  if (assocLinkTable[i_assoc_idx].init)
    {
      sctp_initmsg  initm;
      bzero(&initm, sizeof(initm));
      initm.sinit_num_ostreams = assocLinkTable[i_assoc_idx].numStreams;
      initm.sinit_max_instreams =
              hostTable[assocLinkTable[i_assoc_idx].host_index].numInStreams;
      setsockopt( assocLinkTable[i_assoc_idx].sctpInstance ,
```

```
                        IPPROTO_SCTP,
                        SCTP_INITMSG,
                        &initm ,
                        (socklen_t) sizeof(initm)
                        );
        sctp_connectx( assocLinkTable[i_assoc_idx].sctpInstance ,
                        (struct sockaddr *) &assocLinkTable[i_assoc_idx].destIpAddr[0],
                        assocLinkTable[i_assoc_idx].noOfDestinationAddresses ,
                                assocLinkTable[i_assoc_idx].assocSctpId
                        );
        assocLinkTable[i_assoc_idx].state = ASSOC_ACTIVE;
}
```

Figure 8.17 Connect to an SCTP association.

```
else
    {
        assocLinkTable[i_assoc_idx].state = ASSOC_ACTIVE;
        sctp_initmsg   initm;
        bzero(&initm , sizeof(initm));
        initm.sinit_num_ostreams = assocLinkTable[i_assoc_idx].numStreams;
        initm.sinit_max_instreams =
                hostTable[assocLinkTable[i_assoc_idx].host_index].numInStreams;
        setsockopt( assocLinkTable[i_assoc_idx].sctpInstance ,
                        IPPROTO_SCTP,
                        SCTP_INITMSG,
                        &initm ,
                        (socklen_t) sizeof(initm)
                        );
        /* listen for any possible incoming associations */
        listen( assocLinkTable[i_assoc_idx].sctpInstance ,
                LISTENQ
                );
    }

    return true;
}
```

Figure 8.18 Listen for an SCTP association.

The accept functionality is not really required in SCTP as the communication up indication actually serves that purpose.

The following figure gives an idea of the events which can be received on an SCTP socket. This gives the possibility of handling error situations or reacting to changes in the network. An example is when one of the paths of a multihomed association becomes available.

```
void test_handle_sctp_msg ( int              fd , short int  revents ,
short int *gotEvents ,    void    *dummy )
{
    int                        msg_flags , msg_len ;
    int                        nr_destinations ;
    socklen_t                  peer_addr_len ;
    char                       readbuf[4096];
    struct sockaddr_in         peer_addr ;
    struct sctp_sndrcvinfo     sctp_info ;
    union sctp_notification    *snp ;
    struct sctp_assoc_change   *sac ;
    struct sctp_remote_error   *sre ;
```

```
struct  sctp_send_failed      *ssf;
struct  sctp_shutdown_event   *sse;

int                           support_PRSCTP = 0;


memset(&sctp_info, 0, sizeof(sctp_info));
memset(readbuf, 0, sizeof(readbuf));

peer_addr_len = sizeof(struct sockaddr_in);

msg_len = sctp_recvmsg( fd,
                        readbuf,
                        sizeof(readbuf),
                        (struct sockaddr *) &peer_addr,
                        &peer_addr_len,
                        &sctp_info,
                        &msg_flags
                      );
```

Figure 8.19 SCTP data received.

```
/* figure out if notification or data ?!*/
if (msg_flags & MSG_NOTIFICATION)
  { /* handle SCTP notifications from local peer */
    snp = (union sctp_notification *) readbuf;
    switch (snp->sn_header.sn_type) {
    case SCTP_ASSOC_CHANGE:
      /* SCTP association status changes */
      sac = &snp->sn_assoc_change;
      switch ( sac->sac_state) {
      case SCTP_COMM_UP:
        /* call the normal commUP notification processing */
        test_communicationUpNotif ( sac->sac_assoc_id,
                                    0,
                                    nr_destinations,
                                    sac->sac_inbound_streams,
                                    sac->sac_outbound_streams,
                                    support_PRSCTP,
                                    (void *) &peer_addr
                                  );

        break;
      case SCTP_COMM_LOST:
        /* call the normal commLost notification */
        /* processing */
        test_communicationLostNotif ( sac->sac_assoc_id,
                                      0,
                                      NULL
```

```
                                        );
      break ;
  case  SCTP_RESTART :
    /* call the normal restart notification processing */
    test_restartNotif ( sac ->sac_assoc_id ,
                        NULL
                        );
    break ;
  case  SCTP_SHUTDOWN_COMP :
    /* call the normal shutdown completeed notification */
    /*  processing */
    test_shutdownCompleteNotif ( sac ->sac_assoc_id ,
                                 NULL
                                 );
    break ;
  default :
    break ;
  }
  break ;
```

Figure 8.20 SCTP notifications for association status changes.

```
case  SCTP_PEER_ADDR_CHANGE:
case  SCTP_REMOTE_ERROR :
  /* sctp send failure event */
  sre  = &snp ->sn_remote_error ;
  break ;
case  SCTP_SEND_FAILED :
  /* sctp send failure event */
  ssf = &snp ->sn_send_failed ;
  /* call the normal sendfailure notification processing */
  test_sendFailureNotif ( ssf ->ssf_assoc_id ,
                          NULL,
                          ssf ->ssf_length ,
                          NULL,
                          NULL
                          );
  break ;
case  SCTP_SHUTDOWN_EVENT :
  /* sctp association shutdown events */
  sse = &snp ->sn_shutdown_event ;
  /* call the normal Peershutdownreceived notification */
  /* processing */
  test_peerShutdownReceivedNotif ( sse ->sse_assoc_id ,
                                   NULL
                                   );
  break ;
case  SCTP_ADAPTION_INDICATION :
case  SCTP_PARTIAL_DELIVERY_EVENT :
```

```
        break ;
    default :
        cout  <<  "Received_an_unknown_SCTP_notification \n_";
        break ;
    }
}
```

Figure 8.21 SCTP notifications for association events.

```
else  /* handle  normal  user  data  from  remote  peer  */
    {

    test_DistributeMsg ((unsigned int)  sctp_info . sinfo_assoc_id ,
                        (short)  sctp_info . sinfo_stream ,
                        Readbuf ,
                                                    msglen
                    ) ;
    }


}
```

Figure 8.22 SCTP normal data handling.

```
Bool    test_send_msg (int          assoc_id ,
                       data_str     *i_msg_name_ptr ,
                       int          outstream )
{
  int             msg_index ;
  int             result ;
  int             assoc_index ;
  std :: string   temp ;
  std :: string   msg_name ;
  char            temp_char [1000];
  char            temp1 [5];
  int             total_length = 0;
  char            *disp_msg_ptr ;

  /* copy msg into a temp var */
    char* databuf = new char [( sua_message_table [ msg_index ]. length + 1)];
    memcpy ( (char *) databuf ,
            &sua_message_table [ msg_index ]. contents ,
            sua_message_table [ msg_index ]. length
            ) ;

    // send message
    struct sctp_sndrcvinfo   evnts ;
    struct sctp_sndrcvinfo   *sinfo ;

    sinfo = (struct sctp_sndrcvinfo *) &evnts ;

    memset( sinfo , 0, sizeof(struct sctp_sndrcvinfo ));

    sinfo ->sinfo_assoc_id = assocLinkTable [ assoc_index ]. assocSctpId ;
    sinfo ->sinfo_ppid = htonl (SUA_PPI);
    sinfo ->sinfo_stream = outstream ;
    sinfo ->sinfo_flags = 0;

    result = sctp_send ( assocLinkTable [ assoc_index ]. sctpInstance ,
                        (void *) databuf ,
                        sua_message_table [ msg_index ]. length ,
```

```
                                sinfo ,
                                MSG_WAITALL
                          );

       /* delete the temporary buffer */
       delete databuf;

    return true ;
}
```

Figure 8.23 SCTP data send function.

8.5   Stream-Based Communication

Normal 2 way stream association

TCP Applications can convert to the normal model of an SCTP stream-based association.

TCP application can use the same mode of operations in SCTP. A stream of bytes can be sent to the remote peer or a stream of byte can be received from the remote peer. As the transport layer does not delineate the information within the stream, it is completely up to the application to extract the information from the stream. An element extracted from the stream does not arrive in one go but can in theory trickle into the application one byte at a time. This means that applications have to employ their own buffer management for extracting the relevant information from the stream. This also makes it harder to detect overflow of the data elements within the stream.

Most applications try to constrain the duration for which the association is used.

Advanced 2 way stream association

This employs all the transport layer indications coming up from the SCTP transport layer.

Most of an SCTP stream application is similar to the already described SCTP message-based communication.

The biggest difference is in how the data within association are received. The application receives a message with a certain data length, but has to reassemble the received message to figure out if any data elements are contained within the segments. In theory, data over a stream-based SCTP association can be received one byte at a time until the application has determined that it has received a complete data element.

8.6   Secure SCTP Association

Most applications are required to have a secure variant of the association. For some applications an IPSEC tunnel is used for communication between the 2 peers. As IPSEC is a network-based security, it is not covered here. Other applications use transport-based communication. HTTP is the unsecured web protocol while HTTPS is the secured version of the web protocol. The application can use the same secure port as with TCP. The security protocol used is Transport Layer Security (TLS).

It can provide security for both datagram and stream-based communication. The application will use the API provided by the TLS located above the SCTP. Thus the TLS version of the application protocol does NOT use the socket API as shown in above paragraphs.

The best known TLS library is openSSL (OpenSSL).

The layout of code for communication is similar to the nonsecure SCTP association but some specific security functionality is added.

Create SSL Context

ctx = SSL_CTX_new(DTLSv1_server_method());
This function sets up the SSL context(ctx) of the association.

Setup SSL Context Cipher list

SSL_CTX_set_cipher_list(hostTable[i_host_idx].ctx, "ALL:NULL:eNULL:aNULL");
This function sets up a list of ciphers to be used by the SSL context.

Setup SSL Context verifycation

SSL_CTX_set_verify(hostTable[i_host_idx].ctx, SSL_VERIFY_PEER |
SSL_VERIFY_CLIENT_ONCE, dtls_verify_callback);
This function verifies that the client authenticates the SSL context.

Setup SSL Context Read Ahead

SSL_CTX_set_read_ahead(hostTable[i_host_idx].ctx, 1);
This function reads the SSL context when a message is received in the SCTP association.

The secure SCTP socket needs to be created as in previous examples. Some extra work is needed to set up the secure association.

```
bool   test_create_SCTP_socket_SSL(int i_host_idx)
{
    struct sctp_event_subscribe  events;
    unsigned int i;
    int    res;
    int    pid;
    SSL_CTX *ctx;
    struct pass_info *info;
    const int on = 1;

OpenSSL_add_ssl_algorithms();
SSL_load_error_strings();
hostTable[i_host_idx].ctx = SSL_CTX_new(DTLSv1_server_method());
SSL_CTX_set_cipher_list(hostTable[i_host_idx].ctx, "ALL:NULL:eNULL:aNULL");
pid = getpid();
if(!SSL_CTX_set_session_id_context(hostTable[i_host_idx].ctx, (void*)&pid,
    sizeof pid))
        /* SSL_CTX_set_session_id_context */
```

```
     return false;

if (!SSL_CTX_use_certificate_file(hostTable[i_host_idx].ctx, "server-cert.pem",
     SSL_FILETYPE_PEM))
          /* ERROR: no certificate found */
     return false;

if (!SSL_CTX_use_PrivateKey_file(hostTable[i_host_idx].ctx, "server-key.pem",
     SSL_FILETYPE_PEM))
          /*ERROR: no private key found */
return false;

if (!SSL_CTX_check_private_key (hostTable[i_host_idx].ctx))
/* ERROR: invalid private key! */
          return false;

/* Client has to authenticate */
SSL_CTX_set_verify(hostTable[i_host_idx].ctx, SSL_VERIFY_PEER |
     SSL_VERIFY_CLIENT_ONCE, dtls_verify_callback);

SSL_CTX_set_read_ahead(hostTable[i_host_idx].ctx, 1);
```

Figure 8.24 Create SSL session.

```
/* open the SCTP socket and get the filedescriptor */
hostTable[i_host_idx].sctpInstance = socket( AF_INET,
                                             SOCK_SEQPACKET,
                                             IPPROTO_SCTP
                                           );

if (hostTable[i_host_idx].sctpInstance < 0)
  {
     return false;
  }

setsockopt(hostTable[i_host_idx].sctpInstance, SOL_SOCKET,
     SO_REUSEADDR,  (const void*)&on, (socklen_t)sizeof(on));

/* set the sort of SCTP notifications that test application is interested in */
bzero(&events, sizeof(events));
events.sctp_data_io_event = 1;
events.sctp_association_event = 1;
events.sctp_shutdown_event = 1;
res = setsockopt( hostTable[i_host_idx].sctpInstance ,
                 IPPROTO_SCTP,
                 SCTP_EVENTS,
                 &events,
                 sizeof(events)
               );

/* Create BIO to set all necessary parameters for  following connections, */
/* e.g. SCTP-AUTH         */
BIO_new_dgram_sctp(hostTable[i_host_idx].sctpInstance, BIO_NOCLOSE);
```

Figure 8.25 Create SCTP socket and link to SSL session.

```
/* register the test callback functions for the SCTP filedescriptor */
/* so if a msg is received by SCTP, it can be handled by the */
/* test callback function */
res = test_register_fd( hostTable[i_host_idx].sctpInstance ,
                        EVENT_TYPE_SCTP,
                        POLLIN|POLLPRI,
                        (void *) &test_handle_sctp_msg ,
                        NULL
                      );
if (res < 0)
```

```
        {
          return false;
        }

   for (i=0; i< hostTable[i_host_idx].numLocalIpAddrs; i++)
      {
        /* bind each address to the socket */
        res = sctp_bindx( hostTable[i_host_idx].sctpInstance,
                          (struct sockaddr *) &hostTable[i_host_idx].localIpAddr[i],
                          1,
                          SCTP_BINDX_ADD_ADDR
                        );
        if ( res != 0)
          {
            // not success
            return false;
          }
      }
}
```

Figure 8.26 Register SCTP SSL socket and bind addresses to it.

It is somewhat more complicated to accept or connect a secure SCTP association.

```
Bool   connect_accept_assoc_SSL (int i_assoc_idx)
{
   //association ID not avalaible
   assocLinkTable[i_assoc_idx].assocSctpId = -1;

   if (assocLinkTable[i_assoc_idx].init)
     {
        sctp_initmsg   initm;
        bzero(&initm, sizeof(initm));
        initm.sinit_num_ostreams = assocLinkTable[i_assoc_idx].numStreams;
        initm.sinit_max_instreams =
              hostTable[assocLinkTable[i_assoc_idx].host_index].numInStreams;
        setsockopt( assocLinkTable[i_assoc_idx].sctpInstance,
                    IPPROTO_SCTP,
                    SCTP_INITMSG,
                    &initm,
                    (socklen_t) sizeof(initm)
                  );

        OpenSSL_add_ssl_algorithms ();
        SSL_load_error_strings ();
        ctx = SSL_CTX_new(DTLSv1_client_method ());
        SSL_CTX_set_cipher_list(assocLinkTable[i_assoc_idx].ctx, "eNULL:!MD5");

        if (!SSL_CTX_use_certificate_file(assocLinkTable[i_assoc_idx].ctx,
             "client-cert.pem", SSL_FILETYPE_PEM))
           /* ERROR: no certificate found */
     return false;

        if (!SSL_CTX_use_PrivateKey_file(assocLinkTable[i_assoc_idx].ctx,
             "client-key.pem", SSL_FILETYPE_PEM))
           /* ERROR: no private key found */
   return false;

        if (!SSL_CTX_check_private_key (assocLinkTable[i_assoc_idx].ctx))
           /* ERROR: invalid private key */
   return false;

        SSL_CTX_set_verify_depth (assocLinkTable[i_assoc_idx].ctx, 2);
        SSL_CTX_set_read_ahead(assocLinkTable[i_assoc_idx].ctx,1);

        assocLinkTable[i_assoc_idx].ssl = SSL_new(assocLinkTable[i_assoc_idx].ctx);
```

```
                /* Create DTLS/SCTP BIO and connect */
                assocLinkTable[i_assoc_idx].bio =
                   BIO_new_dgram_sctp(assocLinkTable[i_assoc_idx].sctpInstance, BIO_CLOSE);

                sctp_connectx( assocLinkTable[i_assoc_idx].sctpInstance,
                               (struct sockaddr *) &assocLinkTable[i_assoc_idx].destIpAddr[0],
                               assocLinkTable[i_assoc_idx].noOfDestinationAddresses,
                                        assocLinkTable[i_assoc_idx].assocSctpId
                            );

                 SSL_set_bio(assocLinkTable[i_assoc_idx].ssl,
                      assocLinkTable[i_assoc_idx].bio, assocLinkTable[i_assoc_idx].bio);
                 /* handle the SCTP notifications */
                 BIO_dgram_sctp_notification_cb(assocLinkTable[i_assoc_idx].bio,
                      &handle_notifications, (void*) assocLinkTable[i_assoc_idx].ssl);

                if (SSL_connect(assocLinkTable[i_assoc_idx].ssl) < 0)
                {
                        /* SSL_connect error */
                        printf("%s\n", ERR_error_string(ERR_get_error(), buf));
                        exit(-1);
                }

                assocLinkTable[i_assoc_idx].state = ASSOC_ACTIVE;
```

Figure 8.27 Connect an SCTP SSL association.

```
        }
    else
       {
          assocLinkTable[i_assoc_idx].state = ASSOC_ACTIVE;
          sctp_initmsg   initm;
          bzero(&initm, sizeof(initm));
          initm.sinit_num_ostreams = assocLinkTable[i_assoc_idx].numStreams;
          initm.sinit_max_instreams =
             hostTable[assocLinkTable[i_assoc_idx].host_index].numInStreams;
          setsockopt( assocLinkTable[i_assoc_idx].sctpInstance,
                      IPPROTO_SCTP,
                      SCTP_INITMSG,
                      &initm,
                      (socklen_t) sizeof(initm)
                      );
          /* listen for any possible incoming associations */
          listen( assocLinkTable[i_assoc_idx].sctpInstance,
                  LISTENQ
                );
       }

    return true;
}
```

Figure 8.28 Listen on an SCTP SSL association.

After setting up the association, data can be send to the remote peer.

```
Bool   test_send_msg_SSL(int               assoc_id,
                         data_str     *i_msg_name_ptr,
                         int          outstream)
{
   int          msg_index;
   int          result;
   int          assoc_index;
   std::string  temp;
   std::string  msg_name;
   char         temp_char[1000];
   char         temp1[5];
```

```
int              total_length = 0;
char             *disp_msg_ptr;

/* copy msg into a temp var */
    char* databuf = new char[(sua_message_table[msg_index].length + 1)];
    memcpy ( (char *) databuf,
              &sua_message_table[msg_index].contents,
              sua_message_table[msg_index].length
            );

    struct sctp_sndrcvinfo   evnts;
    struct sctp_sndrcvinfo   *sinfo;
    sinfo = (struct sctp_sndrcvinfo *) &evnts;

    // send message
    len = SSL_write(assocLinkTable[assoc_index].ssl,
                              databuf,
                              test_message_table[msg_index].length);

    switch (SSL_get_error(ssl, len))
    {
    case SSL_ERROR_NONE:
        BIO_ctrl(assocLinkTable[assoc_index].bio,
                          BIO_CTRL_DGRAM_SCTP_GET_SNDINFO,
                          sizeof(struct sctp_sndrcvinfo), &sinfo);
        printf("wrote_%d_bytes ,_stream:_%d,_ssn:_%d,_ppid:_%d,_tsn:_%d\n",
                  (int) len, (int) sinfo.sinfo_stream, (int) sinfo.sinfo_ssn,
                  (int) sinfo.sinfo_ppid, (int) sinfo.sinfo_tsn);
        break;
    case SSL_ERROR_WANT_WRITE:
        /* Just try again later */
        break;
    case SSL_ERROR_WANT_READ:
        /* continue with reading */
        break;
    case SSL_ERROR_SYSCALL:
            perror("Socket_write_error");
            exit(1);
            break;
    case SSL_ERROR_SSL:
        printf("SSL_write_error:_");
            printf("%s_(%d)\n", ERR_error_string(ERR_get_error(), buf),
                  SSL_get_error(ssl, len));
            exit(1);
                              break;
    default:
            printf("Unexpected_error_while_writing!\n");
            exit(1);
        break;
    }

    /* delete the temporary buffer */
    delete databuf;

    return true;
}
```

Figure 8.29 Send message over SCTP SSL association.

Similar to the above-mentioned receiveMsg functions, the message can be received from the socket, but the actual handling of the SSSL session requires some extra work.

```
void test_handle_sctp_msg_SSL(int fd, short int revents, short int *gotEvents,
void *dummy)
{
```

```
int                              msg_flags , msg_len ;
int                              nr_destinations ;
socklen_t                        peer_addr_len ;
char                             readbuf [4096] ;
struct sctp_sndrcvinfo           sctp_info ;

memset(& sctp_info , 0 , sizeof ( sctp_info ));
memset ( readbuf , 0 , sizeof ( readbuf ));

msg_len = SSL_read ( assocLinkTable [ assoc_index ] . ssl ,
                                    readbuf ,
                                    sizeof ( readbuf ));

switch ( SSL_get_error ( assocLinkTable [ assoc_index ] . ssl , len ))
{
  case SSL_ERROR_NONE :
      BIO_ctrl ( assocLinkTable [ assoc_index ] . bio ,
                               BIO_CTRL_DGRAM_SCTP_GET_RCVINFO,
                               sizeof ( struct sctp_sndrcvinfo ) , & sctp_info );
      printf ( "read_%d_bytes , _stream :_%d , _ssn :_%d , _ppid :_%d , _tsn :_%d \n" ,
                      ( int ) len , ( int ) sctp_info . sinfo_stream ,
                      ( int ) sctp_info . sinfo_ssn ,
                      ( int ) sctp_info . sinfo_ppid , ( int ) sctp_info . sinfo_tsn );

  test_DistributeMsg( ( unsigned int ) sctp_info . sinfo_assoc_id ,
                      ( short ) sctp_info . sinfo_stream ,
                      Readbuf ,
                                          msglen
                      );

      break ;
  case SSL_ERROR_WANT_READ :
      /* Just try again */
      break ;
  case SSL_ERROR_ZERO_RETURN :
      reading = 0;
      break ;
  case SSL_ERROR_SYSCALL :
              perror ( "Socket_read_error" );
              exit ( 1 );
              break ;
    case SSL_ERROR_SSL :
printf ( "SSL_read_error : _" );
              printf ( "%s_(%d)\n" , ERR_error_string ( ERR_get_error () , buf ) ,
                      SSL_get_error ( ssl , len ));
              exit ( 1 );
      break ;
  default :
               printf ( "Unexpected_error_while_reading !\n" );
                  exit ( 1 );
            break ;
  }


}
```

Figure 8.30 Receive data from an SCTP SSL association.

8.7   How to Use Multihoming

Simple Multihoming

When applications are using the plain SCTP association, it is still possible to utilize multihoming in the association they are setting up with the remote peer. The application is not interested in controlling multihoming itself, only that it can use multihoming without impacting the application logic.

By binding the socket to all or a subset of the IP addresses of the interfaces present in the node, SCTP is able to utilize those interface and multihoming across them. Some systems and network policies set by the network provider are not in favor of allowing this. Care must be taken when doing so.

Difficulties may arise when employing private IPV6 types of addresses because of the limited scope of the addresses within the network. An example are link local addresses, which may be included in the bind ANY to the socket, but when SCTP decides to use the link local address for communication, then no communication will be possible unless the two nodes are part of the same link within the same local network.

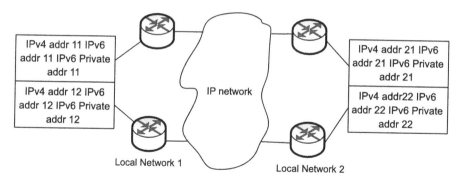

Figure 8.31   General multihomed SCTP association.

IPv6 Private addr11 and IPV6 Private addr12 should not be employed in the multihomed network unless the remote peer host2 is on the same link1 and link2 as the host1. The other addresses in host 1 are all globally routable and can be used for multihoming the SCTP association.

Advanced multihoming

Some applications wish to be informed of events on the multihoming of the SCTP association itself or even to control it themselves. Via the use of SCTP_getpaddrs(), it can obtain all the addresses of the remote peer; the call to getpeername provides the peer address to which the association is presently sending it data. By using the

Link Local Network

Figure 8.32 Multihomed SCTP association with link local addresses.

sctp_sendmsg() and providing a specific address, the application can override the primary address and send data on the application specific address. If a remote peer address is not part of the association, then an SCTP_sendmsg will result in setting up a new association.

SCTP can generate a notification which is relevant to multihoming: SCTP_PEER_ADDR_CHANGE. One of the following state values may be present:

1. SCTP_ADDR_ADDED: an address has been added to the SCTP association

2. SCTP_ADDR_AVIALABLE: the address is now reachable

3. SCTP_ADDR_CONFIRMED: the address has been confirmed and is valid

4. SCTP_ADDR_MADE_PRIM: the address has now been made the primary address within the association (different SCTP associations may have different primary addresses towards the same node, as long as any of the addresses are reachable)

5. SCTP_ADDR_REMOVED: the address is no longer part of the association and has been removed from the association

6. SCTP_ADDR_UNREACHABLE: the address can no longer be reached, and is still part of the association but no traffic can be exchanged utilizing the address

In order to obtain these notifications in good order, it is required that SCTP be doing its heartbeat mechanism toward the remote peer. This is enabled by default (even for simple SCTP use with multihomed hosts). The threshold can be set via the setsockopt() for the number of missed heartbeats or the number of retransmission timeouts before a destination address within an association becomes unreachable.

Bibliography

OpenSSL (n.d.) *OpenSSL Project*. Retrieved from OpenSSL - Cryptography and SSL/TLS Toolkit: http://www.openssl.org

Stevens, W 2004 *Unix Network Programming the Sockets Networking API*. Addison-Wesley

Stewart RR and Xie QX 2001 *Stream Control Transmission Protocol (SCTP) a Reference Guide*. Addison-Wesley, New York, NY

Tüxen, M (n.d.) *www.SCTP.de*. Retrieved from SCTP: http://www.sctp.de

Index